For Colt,
I hope
ation here -- With Gods blessing,

"NOW, VOYAGER"... A garden speaks of the nature of the gardener; the books in a library tell much about the reader. *God's Opera* reflects the remarkable woman who has given this book to us.

Jettie June Harris at 16 loved walking in the rugged West Virginia hills, brought the most beautiful music from any piano she touched, was transfixed by starlight and sunlight, wonder-filled at the works of Heaven and in the heavens. Closest friends as 16 year olds can be, life's current drew us in many seas until we rejoined a few years ago.

Jettie June now—mea amica aeterna—is still entranced by the mists on the fields and forest of her Lynchburg, Virginia, home; still can bring the most beautiful music from any piano or organ she touches; and is more fearless than ever in her search for truth. A woman of profound faith, she believes that new discoveries about the nature of the universe tell us, if we have ears to hear, messages of confirmation and enrichment.

For example, string theory is among her carefully researched examinations of cutting edge cosmological studies. Strong theory, as she describes it, it a hypothesis that the essential nature of things are infinite numbers of infinitesimally small strings. Some readers may recall Lucretius "De Rerun Natura"[1] which posts an infinite array of infinitesimally small monads. The vibrations of the strings underlie, according to the theory, experienced macro-arrays of matter.

Amazingly, current technology has detected what could be the sound (in our atmosphere) of the strings. It is, beautifully, the pure note of "D." Jettie June has long believed God is, or is expressed as, music and our experienced universe is, in deep space and deep time, music. It follows that understanding music and understanding truth of the Divine are akin, even one.

These conclusions and others reached in *God's Opera* open examinations in theology, music theory, cosmology and astrophysics. To me, as a scientist, "hypothesis" as a best statement consistent with experience and observation is more comfortable language than "proof." In science, a hypothesis may be rejected as implausible; little can be said to be proven. Thus, Jettie June's

Jettie

nexus of faith, science, and music offer worlds of thought into which we can boldly go.

Consider as another example the resonances of her email after a seminar with her Hebrew instructor. This is a woman to whom the Bible means so much, she is learning Hebrew to get closer to the original translations such as the Vulgate or King James' provide. "I learned today," she wrote "that "In the beginning…" really should be translated as "In a beginning…."

"In a beginning…"—were there then other beginnings of which ours is only one? And why not? And what does this "a" mean for how we think about our lives, the world as we know it now, the universe—or the universes—as we know it now?

Readers will find on every page of *God's Opera*, as well as on the pages of her earlier book, *God's Song*, such knowledge, wonder, faith and enriching challenges.

Thank you, Jettie Harris: mea amica aeterna et carissima,

Lois-Ellin Greene Datta

(1) The writing of Lucretius reflect the philosophies and religious mores of the first century BC and are far removed, for the most part, from those of our day. However, it is quite evident from his primary work, De Rerum Natura, *On The Nature of Things*, that the same quests for truth were active those thousands of years ago that we are still grappling with. And what is most amazing is that one of those theories about which Lucretius writes so eloquently is the idea that vibrating strings (strings of something) underlie the "experienced" macro-layers of matter, as this essay reads. No matter how deeply we may disagree with the basic tenets of Lucretius and his theories, there is much to be learned about where we are in the 21st century simply by accepting works such as De Rerum Natura as important contributions to knowledge in general.

JETTIE HARRIS ON ABOVE STATEMENT: The writer of this essay is a learned and highly qualified poet as well as a scientist, and I, as the author of *God's Opera*, respect and cherish her insights and interpretations of those insights. Thank you, as well, mea amica aeterna et carissima.

Endorsements

Jettie Harris is a remarkable person. She has extraordinarily wide-ranging imagination of the significance of music for our lives, drawing on a lifetime of teaching, playing and reading. Those who read this will never hear any music in quite the same way again.

Jeremy Begbie, Duke University, Durham NC

We are learning so much about God's revelation in nature. The fine-tuning of the universe has become a household phrase in science. Whether scientists peer through the telescope or microscope the wonders that are revealed never cease to amaze. Psalm 19 declares that the heavens are chanting the glory of God, and his creation is communicating God's revelation ceaselessly. Jettie Harris' new book takes this seriously and explores the idea that this is more than just abstract truth but that music is central to this revelation. Speaking to Job, the divine voice asks, 'Were you there when the morning stars sang together, when all the angels shouted for joy?' Music is thus present at the beginning of the biblical story, in the middle when the angelic choir announces the birth of Christ, and at the end when heaven is finally united to earth. But maybe the music is playing all the way through the story as well. This book invites us to consider this possibility and stop, look and listen! And in the words of Psalm 19:14, join the choir!

Stephen Dempster, Crandall University, Moncton NB

Since meeting Jettie Harris in 2005 I have been impressed with her love of and knowledge of music and her passion for God and His Word. Her perseverance in pursuing her research on this subject of God, music, and the truth of Genesis about creation is remarkable, especially given the stage of life she is in and often the lack of interest in her work by others. Her hard work has included reading about physics (especially string theory), delivering a paper at the Evangelical Theological Society (a daunting task for someone not trained in theology, especially when, as I recall, less than five people came to her presentation), and studying biblical Hebrew. I commend her intriguing work to thoughtful readers to ponder.

Terry Mortenson, Answers in Genesis, Petersburg, KY

It's been a joy to know Jettie Harris for over six decades from when she was a talented young pianist and singer. Despite more than her share of personal struggles, she worked with noted musicians and seminal thinkers, pursuing every possible angle to unravel the mysterious secret powers of music and how those powers might relate to the very essence of the universe and its creation. An admirable life's journey.

Walt Rybeck, Silver Spring, MD

"...He will rejoice over you with great gladness. With his love, he will calm all your fears. He will exult over you by singing a happy song."

– **Zephaniah 3:17,** *NLT*

God's Opera

FINDING TRUTH IN MUSIC

JETTIE HARRIS

publishers
SOLUTION
The Key to Publishing Possiblities

God's Opera
by Jettie Harris

ISBN-13: 978-0-000000-00-0

publishers SOLUTION
The Key to Publishing Possiblities

Forest, VA
Cover & Interior Design: Megan Whitney
Interior Layout: Christopher Kirk
PublishersSolution.com

Table of Contents

OVERTURE
XI

God's Opera

Table of Contents

God's Opera

Act IV: Music—The Beauty of Holiness

Overture

Characteristically, all musical dramas begin with an "overture." The purpose of the overture is to introduce musical themes, dramatic ideas expressed musically, and less obviously, to set the mood of the performance piece.

The overture is such a definitive piece, in and of itself, that Beethoven eventually went through four different overtures for his opera, Fidelio, before he finally satisfied himself with the accepted version. How well I can understand his dilemma—this book has been on the drawing boards for over forty years if one were to go all the way back to its inception. In fact, it may have its most valid roots in my early childhood when I was literally healed of a dread illness by a piece of music. And, it has suffered somewhat the same sorts of variations and diversions and approaches that Beethoven's overture did.

During my tenth summer, I contracted scarlet fever. The family doctor held out little hope that I would recover. My father bought me a small radio for entertainment, which he set up by my bedside. Every afternoon from 5:15–5:30 P. M., a program aired on WJZ called "The Guiding Light"—it was the story of the Old Testament. The theme music, which introduced and ended the program, was the most beautiful sound I, as a musically aware child, had ever heard. I can recall thinking that I had to be alive the next day to hear it again. Nearly two weeks went by. The fever broke, and my body started to heal. Nobody had any idea where my musical talents came from, but that summer it didn't matter. All that mattered was that it was a piece of classical music, which none of my family or friends had ever heard before that kept me alive until my body could fight off the terrible fever!

This may be a good place to call attention to the immense imagination that I had as a child and to this day still enjoy as a constant presence in my thought life.

As I have aged, I have begun to think of that "imagination," as C.S. Lewis and even Einstein heralded as a tremendous gift to be cherished, in a more serious framework. I call it my "sanctified imagination," and I take serious measures to keep it disciplined for good. I have even gone so far as to give it a name, Jana, a derivative of my own name and to recognize it as a fine tool for meditating on the things that puzzle me until I can resolve them. In that light, consider that the purpose of this "overture" is parallel to the purpose of the operatic overture. The general outline of my thinking about music and creation, theology and music, science and theology and the surprising function of language as a cohesive element will all be explored. And I invite the reader to engage his own imagination in the process.

The most relevant idea to be examined is:

Overture

WHY THE OPERA AS A TEMPLATE?

The use of this art form as an analogy for the creation chronicles may seem odd at first.

Using opera as a template was deliberate on my part. But it must be understood that it has nothing to do with the philosophical, political or theological frameworks which were so evident in the mid- to late-19th century when opera as an art-form was burgeoning. My choice of this art-form is quite simple. It is prompted by the philosophy behind Richard Wagner's expansive re-working of the opera format in order to provide a more satisfactory and useful vehicle for his huge imagination and the demands that this imagination put on opera as it stood in the early years of his artistic growth: as patriarch of the opera in 19th century Europe, he changed the landscape of opera composition and performance forever.

"Wagner's main philosophical contribution to opera was the idea of the Gesamtkunstwerk, or "total work of art." In articles he wrote while in exile, he explained that in opera, the orchestra, singers, text, and scenery were all crucial elements of the total artistic impression and should therefore be given equal weight. What this meant in practice was that the orchestra took on an even greater role than before, and the musical flow became completely uninterrupted from the beginning to the end of an act. This also meant that the aria, traditionally a showpiece for the singer, practically disappeared to be replaced by a more constant stream of musical ideas.

To give order to this new concept of opera, Wagner invented the *leitmotif*, which is a relatively short melodic idea that represents a certain character, thing, or idea. A *leitmotif* associated with a particular character often appears in the orchestra while being mentioned by another character, effectively acting as a stand-in. More complex *leitmotifs* include

the "Tristan chord," a particularly memorable harmonic progression in *Tristan und Isolde* first heard in the prelude (or overture) which comes to symbolize the love and death of the two title characters. "While Wagner never used the word *leitmotif* to describe his own operas, it is a clearly discernible technique in his music which has influenced many other composers. It has been recognizably used since then in film scores."[1]

There are two specific ideas proffered by Wagner's thinking which need to be examined: 1) "total work of art"—in its most idealistic form, the opera gives every facet of the work equal place and equal importance. In other words, singer, orchestra, sets, libretto, staging, sound, color, melodic movement, harmony—all are held in equal importance, none are held as more important than another: 2) "leitmotif"—the 'melodic idea' had a specific representation and recurred only to express that representation, whatever it was meant to be. It is understood that everything worked together to construct the whole, but each "leitmotif" had its own unique identity, and, therefore, its own unique purpose and work to do.

If the reader takes just a little time to examine the opera for its real ability to capture and convey the largeness of the event or life it seeks to portray, it may quickly become evident that this is the crown jewel of all musical art forms, even though it may not be the preferred art-form for any given individual. And this need not diminish or negate, in any way, less grand expressions of musical excellence. It is simply that the shape and content of opera by definition embraces many, if not all, other musical and artistic forms in one way or another. In addition, the genesis of the art form as a concept is extremely interesting. Consider the following definition:

1   http://www.conservapedia.com/Richard_Wagner

opera (n.) "a drama sung"[2] [Klein], 1640s, from Italian **opera**, literally, "a work, labor, composition," from Latin **opera** "work, effort" (Latin plural regarded as feminine singular), secondary (abstract) noun from "**operari**" to work," from **opus** (genitive **operis**) " a work" (see **opus**). Defined in "Elson's Music Dictionary" as, "a form of musical composition evolved shortly before 1600 by some enthusiastic Florentine amateurs who sought to bring back the Greek plays to the modern stage."[3]

It is interesting to take note of the etymology of the word "opera" above. It will be obvious as this writing progresses that one of the driving energies behind my entire journey has been the understanding that everything in the universe has a unifying common thread running through it. Of course, from my perspective, that common thread is to be expected because it is the complete expression of the Omniscient Creator, God Almighty.

Literally, the word "opera" means work, effort, and emanates from the concept of doing "work." The genitive form of the word, being "the act of possession," infers that the Creator also possesses that which He "worked" to accomplish. The creation, then, is "His." This idea is in complete agreement with the Biblical concept of God the Father as being the Owner of all creation from conception to consummation at the end of time when we shall finally see "as we are seen" and understand "as we are understood."

---

2 My Note: The definition, "a drama sung," was also especially interesting since the underlying tenet of my thinking about creation is that it **was** a musical event. The vastness and analogical properties of the opera seem to co-relate with the creation saga in many ways.

3 On-Line Etymology Dictionary: 2001-2014: Douglas Harper

Comparing opera to work and to creation, then, takes on a deeper meaning. It is not simply a good analogy: the idea of "opera" becomes the framework for understanding the narrative of the creation chronicle as a musical drama, with a body of musical building blocks—with sound and vibration being the key elements—as the raw materials out of which that drama is constructed and sustained. And it is also, then, at this juncture, to perceive creation as what it really is—the work or Word of God. An evasive concept.

Another, perhaps less profound analogy is the completeness of the opera in terms of embracing so many other art forms: **visual arts** of all kinds from sculpture to painting to scenic transport, **dance, literature,** because most, if not all, operas tell a story, **fashion, architecture, all of the theater arts** from **acting to mime to fantasy and from comedy to tragedy.** And, not least of all, the actual **musical beauty** of the opera itself. In my mind, no other art form provides as complete a metaphor aimed at portraying or exemplifying the drama of creation.

It is interesting to note that the folkloric literature and art of many ancient and lost cultures display similar versions and explanations of the creation chronicle. The Navajo, the Seminole, the Arapaho, the Aboriginal people of Australia and New Zealand, along with many others, all have a trove of wonderful stories telling of their singing creator-gods who brought all that **is** into being by the power of their own thoughts and songs. And these are simply characteristic examples among many more, but the theme is the same and very much in tune with the imaginations of Tolkien and Lewis. What must be noted is the fact that, although totally diverse and spread over a sizable span of history, these collections of lore and art are amazingly similar and carry the same seed of creative energy—that of a musical, singing creator.

Any attempt to deal with musical issues of any kind is a tremendously difficult task. If there are a hundred people exchanging ideas about music or musical preferences, it is a given that there will be a hundred diverse ideas or preferences expressed. It has been said that there is no more subjective art-form than music. Therefore, because of the vastness and inherent differences in the basic understanding of music and its multiplicity, it is necessary to clarify my approach to this idea that musical elements were likely used as the raw materials of creation before launching into the hypothesis itself.

First of all, one thing must be made quite clear. **In no wise** am I claiming that music as we know it in the 21st century is the stuff of creation. Quite to the contrary. As I see it, music as we know it is more **the result** of the inherency of music in all things than it is **the outcome** of a human activity. There are eight basic elements extant in the composition and performance of musical expression: rhythm, harmony, melody, dynamics, texture, tone color, form and, sometimes, text. While these elements are fairly consistent in all musical genre to one degree or another, they are, after all, the result of man's analytical and developmental skills as he gradually discovered their existence in an already well-ordered creation construct.

What, then, am I referring to when I say that the raw materials employed by the Creator at the moment of universal creation were musical elements and that the creative building blocks were unequivocally musical in nature? Dr. Patricia Gray of the School of Music, Theater and Dance, University of North Carolina Greensboro has inferred as a result of her own research that **there may be forms of music that we simply do not know about yet**. She may have been referring to particular genre or

styles of music, but this statement can just as easily extend itself to musical qualities and definitions as yet undiscovered.

Einstein was probably the first to articulate the belief that the universe had at its core a magnificent simplicity—that the elements of the universe were grandly made up of some single, irreducible something that would explain all the wonders that were so elegantly and beautifully evident. Brian Greene states Einstein's premise very clearly:

> "Albert Einstein became obsessed early on by searching for a means of 'describing nature's forces within a single, all-encompassing, coherent framework—he was driven by a passionate belief that the deepest understanding of the universe would reveal its truest wonder: the simplicity and power of the principles on which it is based. (He) wanted to illuminate the workings of the universe with a clarity never before achieved, allowing us all to stand in awe of its sheer beauty and elegance.'"[4]

There is a fundamental confidence in my being by this time **that:** 1) Creation is a thoroughly thought-out and orchestrated event; 2) Creation of anything—and most certainly of the very universe itself which is our bedrock—must have raw materials chosen by God's ordinance; 3) Music—and all of its elements and rudimentary characteristics and data— is offering itself as the one thing that permeates all of life on a high enough plane to comply with the needs set forth by this idea of a concise, beautiful and "elegant universe."[5]

---

4   *The Elegant Universe*: Brian Greene: W. W. Norton & Company, Inc. 1999:Preface, Pg. xi.

5   Ibid

> *"I am enough of an artist to draw freely upon my imagination. Imagination is more important than knowledge. For knowledge is limited, whereas imagination embraces the entire world, stimulating progress, giving birth to evolution. It is, strictly speaking, a real factor in scientific research."*
>
> —*Albert Einstein*

JANA & HER PLACE IN THE JOURNEY TOWARD TRUTH

God's Song[6] was the story of my life and my first faltering steps into a world where truth was exquisitely necessary to the entire chronicle of the creation of the universe and all that is contained within it. The ideas and theses set forth in *God's Song* are more fully developed in this second book, supported by the advancing maturity of a decade of time since the early attempts at expressing such new and different thinking. In that first book, Jana was my imaginary friend and alter-ego who did a great job of conveying my thoughts, and I would prefer to write her into my thinking again, even though the ideas and theses are well-developed by this time and are meant to be taken very seriously. Jana in no way infers a fictional story, but rather affords the opportunity for dialogue and discussion to aid in the process of conveying what I believe. On the other hand, I fully believe that she will lend an air of reality and maybe even fun and excitement to the more serious ideas that I wish to convey. Readers may find a

---

6 *God's Song* 1st edition: Liberty University Press: Lynchburg VA: 2010; *God's Song*, revised 2nd edition, Publishers Solution: Forest, VA: 2017. This book is the "pre-quel" to *God's Opera*.

person in Jana to whom they can relate in a more realistic way than they could relate to theories and abstract suppositions. If they can join her in 'sanctified imaginings' instead of trying to delve into the subtleties of string theory or cymatics, it would be a very good thing.

I have made a life-long mockery of the fact that the only time I can think clearly is when I am talking. To have a companion, even an imaginary one, with whom to carry on discussions is a great gift for one who is so attuned to the spoken word when accuracy and clarity are so vital. Therefore, Jana would be a great help to my personal endeavors in the writing of this book. To further clarify the reason for taking Jana into this writing with me, it might be enlightening to re-tell the story of a little three-or-four-year-old girl in Texas whose best friends were three pilots born only in her imagination. I was that little girl.

It was probably 1934 or 35. Maybe 1936. At any rate, I was very young. My father was a lover of airplanes and often drove the family down to Amarillo on Sunday afternoons to watch them taking off and landing. One Sunday afternoon, a strange-looking aircraft landed and a beautiful, young female pilot dismounted. The aircraft was huge. It had normal wings and engine but it also had four giant blades like giant propellers over the top of the cockpit which revolved slowly and looked similar to our present-day helicopters. I had never seen anything like it before. All I knew was that it was BIG. The pretty pilot had curly hair, a space between her teeth, and was wearing a leather jacket, helmet and goggles.

After speaking to the line-boys, she walked toward my mother and father and me, and as she passed near us, she stopped, came over to where we were standing, ruffled my curly hair and said, 'Little girl, one day you will fly one of these. Would you like that?' I was stunned. Of course, that was Amelia Earhart!

My father read the lettering on the side of the funny airplane to me: "Pitcairn Autogiro." He remarked that it was very 'muddy.' Probably because it had been landing in muddy fields all across the nation.

Who knows how things register in a child's mind? How many little boys have called a 'caboose' a 'cuss-buss' or a 'choo-choo-train' a 'gnk-gnk-chane'? I suppose that somewhere in my childish mind there was a connection between 'auto-giro' and 'Genery.' Several weeks later, my mother noticed me out in the backyard on the swing talking loudly to 'Pit,' 'Genery' and 'Mud.' For the next five or six years, Pit, Genery and Mud and I travelled all over the world in the 'big plane with the funny propeller'—Australia, Alaska, Russia, Germany. We would crash, never get hurt and go on to another adventure. Obviously, my three friends were the result of my encounter with the *Pitcairn Auto-giro covered with mud*. They also were, no doubt, the beginning of my love affair with aviation and the empowerment behind my learning to fly.

> *"Logic will get you from A to Z; imagination will get you everywhere."*
>
> —*Albert Einstein*

JANA'S IMAGINATION DEALS WITH WORDS

This age of technological explosion has changed so much of our lives—much of it without our even being aware of the changes. In my youth, words and writing and literature were the means for so much pleasure. In fact, to sit at one's desk and pen imaginary stories, or write a love letter, or to make an entry in a journal was an exquisite experience of self-expression, which brought a sense of gratification to the

writer, as well as the possible recipient of the penned word. One of the most prized possessions that I ever had was a poem written by hand about me by a friend who has remained precious to me for over sixty years. I still revel in the emails I receive from this friend who now lives half-way around the world from me because her prowess with the written word is so beautiful and so far more developed than most people could ever imagine.

That gift of being intimate with words expresses real beauty and I can't help but think that it is also warp and woof of the very heart of my own theory of creation—words are living, dynamic things—spoken or sung. There is, as demonstrated in the Hebrew language, a very close relationship between the words which translate as speaking and singing. Also, among conservative Gentiles, there is a strong sentiment that supports the idea that the first language was, indeed, the language of God Himself as He began to communicate with His people, the Israelites who became a people, by His own decree, through the lineage of Abraham and Sarah. That language is Hebrew—Biblical Hebrew.

Even as a young child I was interested in words. Where they came from and why they meant what they meant. Who made them that way? Why were words in languages that I could not understand so much like the words that I knew in English—because many of them were very much alike. Then I noticed that the preacher was always talking about the Word of God and I wondered why God would have a special Word? Later I found out that the preacher meant that Jesus was that Word and that was quite confusing. Until I was much older and began to understand that all words are dynamic and not static. They live and breathe and move and have power, for good or for bad. And this Person called the Word of God was Jesus Christ, the living personification of God, the God Who could

not be seen except in the Person of Jesus. That remained a mystery which I could not fathom for many years.

Then as a much older person, when I studied Hebrew and found out that words to the Jewish mind were even more alive and had even greater power, I was really intrigued. All of this goes into the making of a truly magnificent enigma: The energy that makes words so powerful is a creative energy, and it really does have something to do with the Word of God Who is Co-Creator with God the Father.

For the casual reader who may be grappling with such evasive ideas as the dynamism of words and the creative energy held by them as it may relate to the person of Jesus, the Co-Creator, Who is called by the name 'Word of God,' I would say: Befriend your enigmatic person, your Jana, and allow him/her to pique your imagination and try exploring new worlds which you may never have visited before. You may find, to your great surprise, new vistas of understanding about music, about creation, about yourself.

I find it strangely un-nerving to watch the unfolding of texting-language, which is largely a language of abbreviations, in the 21st century. How sad. In the first place, the words themselves lose their identity to a series of initials. The words are never spoken—only their acronyms are understood to have meaning. And just imagine—any depth of emotion is wasted on short-speak streaming across a screen in place of the musical rise and fall of the human voice as it frames words of love—or hate—or empathy—or disdain—or fear—or victory.

Without this grasp of the absolute necessity for words in the fully developed and mature human being's sphere of life and influence, it would be quite difficult to make that final journey into a world that in all proba-

bility is not only created by "**words**"—sung or spoken—but is also sustained and maintained by the energy of "**words**." Words are the essence of sound. The sung word is the expanded essence of sound. The founders of the emerging science of cymatics make the bold statement that without sound there would be no form. They express it this way:

> "Vibration underpins all matter in the universe. No matter
> can exist without sound and vibration."[7]

When Jana ponders ideas like this, it is not a matter of consternation to her, because being of an imaginative nature and being, she can travel into all kinds of places in thought and can allow those thoughts to just **be** while she reviews them—without constraint.

It is my plan to shape this book, *God's Opera,* loosely as a dialogue between my own discoveries and thoughts and the imaginative ruminations of Jana, as my alter-ego, or other "version of myself." The reason for this is to illustrate the process by which I came to the final formulation of my theory of the "Singing God and Creation" as a musical phenomenon over a life-time of imaginative intellectual roamings, and to give a frame-work to the ideas that eventually led me to the concept of music as the raw material of creation itself. It is also my intent to give the reader a vehicle for the same kind of intellectual roaming when the ideas themselves may become difficult to deal with in the accepted manner of 21st-century scientific empirical thinking.

> *"Imagination is everything. It is the preview of life's coming attractions."*
>
> —*Albert Einstein*

My sanctified imagination, which I am now personifying as "Jana," helped me through the maze of finding logical support for my theory in the demanding worlds of science, language, art and theology. It is not difficult for science, for instance, to say that without sound, matter could not exist. But to take that last elusive, uncharted leap into a world which can say with conviction that the sound they are talking about is in reality music and that this sound—this music—has its genesis in the Word (Song) of God, becomes a very difficult issue. It is difficult because of the very ubiquitous nature of music. It is difficult because of the infinite, eternal nature of God. Furthermore, to understand that the music of creation probably has very little semblance to the thing that we humans now call music is problematic without engaging the ability to use your 'sanctified imagination' and draw the contrast between what humans create and the eternal and utter beauty of what God intends.

It has been an amazing journey and with the presence of Jana as the curator of the sanctified imaginations that were necessary to the process, I hope to introduce the reader to this exciting world of possibilities.

BACK TO THE OPERA

I remember thinking as a young musician that operas were unending and that was one of the reasons that I did not develop a love for opera in my youth. The irony is that it actually is that particular characteristic of opera—its longevity and its all-encompassing nature—that makes it the **perfect foil** for explaining the hugeness of the creation chronicle.

Wikipedia makes an interesting observation about the word "foil:"

"The word foil comes from the old practice of backing gems with foil in order to make them shine more brightly."[8]

The use of the "opera" as an analogy was serendipitous to the writing of this book. And it was also serendipitous that I stumbled onto the archaic expression "perfect foil" while looking for ideas for comparison. This antique setting of the word "foil" has nothing to do with the idea of "thwarting." The idea of a **backing** being used to make the creation story shine more brightly for the average reader is another marvelous bit of beautiful irony. And, who knows, use of the "perfect foil" as an archaic word picture—a "shiny backing"—may supply just that extra dimension of newness to the story so oft-told to engage others in becoming involved in the quest for "truth in music" in terms of the creation saga. And, most certainly, newness in thinking of Almighty God in His great Trinity, as the "Singing God" Whose voice *is and was and ever shall be* the vibrating energy—Whose song (sound) created and still maintains all of the vast cosmos which so evades our understanding and the understanding of our greatest thinkers and scholars.

At the beginning of this "Overture," I defined Wagner's concept of opera as "total work of art." This epitomizes my own concept of the enormity and yet simplicity of everything that "is." I repeatedly return to the idea that Einstein spent the last thirty or forty years of his life persistently pursuing: trying to describe "nature's forces within a single all-encompass-

---

8   Wikipedia® is a registered trademark of the Wikimedia Foundation, Inc.

ing, coherent framework—the simplicity and power of the principles on which it (nature's force—the universe) is based."[9]

The paradox of opera makes it possible to talk about the paradoxes of creation. In 2013 the Metropolitan Opera mounted a new production of Wagner's *Die Walkure*. The Met's online newsletter had this to say:

> "The second installment of Robert Lepage's new production of the *Ring* cycle conducted by Fabio Luisi, features a stellar cast led by Bryn Terfel as Wotan, lord of the Gods, and Deborah Voigt as Brünnhilde. Jonas Kaufmann and Eva-Maria Westbroek star as the Wälsungen twins, Siegmund and Sieglinde, and Stephanie Blythe is Fricka. 'The *Ring* is not just a story or a series of operas, it's a **cosmos**,' says Lepage, …who brings cutting-edge technology and his own visionary imagination to the world's greatest theatrical journey."[10]

The use of the word "cosmos" in describing the Wagner *Ring* cycle is fascinating to me. Apparently, the writer of this article in the Metropolitan Opera newsletter shares the sense that this theatrical production has some of the temporal earmarks of the hugeness which I intend to bring into the consideration of the creation chronicle as a musical event—it is indeed a cosmological event which I believe should be understood as a musical event as well. This is the entire bent of this book—creation as a truly eternal and sacred musical offering of God Almighty to His creatures.

---

9   *The Elegant Universe*: Brian Greene: W. W. Norton & Company, Inc. 1999:Preface, Pg. xi.

10   2014 The Metropolitan Opera

The question must be asked, and will hopefully be answered in the pages of this book: "Why is it necessary to understand creation as a musical offering to mankind, and as a musical event in the total scheme of things?"

It is not necessary to spend time analyzing opera itself. Just to see it as the engulfing art form that it is and to be able to appreciate the grandeur of that art form, will open new vistas of appreciation relevant to a Biblical chronicle which we have heard so often told that it is almost common-place. The marvel is that there is nothing commonplace about it. And the sad realization is, also, that the stark lack of knowledge and insight into the beauty of the opera is a depressing commentary on the decadence of our contemporary view of musical beauty.

I am asking the reader to deliberately lay aside the stereo-typical ideas that may have been so deeply imbedded culturally, if that is the case with any individual reader, and join Jana on an imaginary journey into new places on the operatic stage. It is my hope that each individual reader will simply allow himself to see and hear the immense communicative possibilities of opera as a vehicle for understanding and examining creation as God's ongoing performance of magnificence and ineffable creative power.

It is my intent, as this book progresses, to take various aspects of the opera performance from the temporal, artistic and musical point of view, explain them briefly and use them as a platform for examining whatever aspects of my theory of creation may be in existence at that moment.

For instance, the visual art present in the opera at large, and in the 2013 production of *Die Walkure* in particular, is profound in and of itself. If one were to watch that performance on a computer or television and turn off the sound, the effect would still be astoundingly mesmerizing.

The colors and the lighting and the interplay between them is a captivating spectacle. Think of a sunset by the sea. I lived on an island in New Jersey with the ocean to the east and a bay to the west. One day as I was flying my airplane southward along the seashore, I was amazed to be able to watch as God painted a huge canvas off to my right. The entire panorama of the sky was ablaze with teals and oranges and reds, deep purples and a vivid blue, cloudless sky soaring above this wash of changing colors and a bright dark orange sun sinking silently into the horizon. One could not deny the overwhelming creativity of God Almighty in the reflection of this scene of water and sand and sky. The awesome talent of Robert Lepage's *Ring* cycle productions pale by comparison, but they do provide a fine foil for describing the creativity of God manifest in that sunset. And one does not need to hear the music or the singing in order to understand the beauty of the settings.

And this is only one small component of what makes opera so magnificent. Analogies could (and will) be drawn as this book develops and each facet will be grand in its own right. Even as each individual component part of our glorious universe(s) stands as its own radiant and elegant entity but also exponentially expands the beauty of the whole.

IN SUMMARY
So it is. I shall draw the comparisons as they seem appropriate and meaningful to the unfolding of my theory. And remember, *God's Opera*, the book, is only a "perfect foil" for giving an accessible analogy to the most magnificent of gifts from our Creator God—the creation of our cosmos.

I think the most important facet of this approach to the greatest of all events—creation—is this: the immensity and the complexity and, yet, the utter simplicity and beauty of that creation. It can be studied and

appreciated in all of its unfathomable immensity as the totally immeasurable cosmos that it is; or it can be studied and appreciated in minute increments. But—and ponder this carefully—the more infinitesimal the studies become, the more evident it will also become that there is an untiring unity in it all.

The opera comprises many separate and equally profound and beautiful elements, which when observed each one alone, may seem easy enough to comprehend, but when melded into the whole, the diverse nature of that piece may become overwhelmingly complex. Whether viewed as component parts or seen in complete combination as a musical production, the beauty is never compromised nor rendered questionable. This is, indeed, the composition of our universe. Many interwoven segments comprising the ineffable whole and infusing it with wonderful unity, beauty and elegance.

Weaving this tapestry will require the use of one's "sanctified imagination," as I said earlier on, in order to grasp some of the more abstract or theoretical ideas concurrent with this approach to the creation chronicles. And here is where Jana and her exquisite command of dedicated fantasy will be of great help and, I hope, interest. I shall pose the ideas and then allow Jana to reflect upon them and bring her incisive free-wheeling thought to bear in attempting to put them into perspective or provide some rationale for them.

Jana has had a profound influence in my life from the very beginning. As an infant, I had a stuffed bull-dog named Fido. By the time I was two or so, Fido and I had been to many strange places, most of which I probably did not even actually know about. No matter; if I didn't know where a place was, I would make up a different place and give it a new nonsense-name.

So, it is not surprising that by the time I was in my forties and thinking seriously about the vital beginnings of things, that "sanctified imagination" which I have personified as this imaginary person, Jana, was hard at work in the dim recesses of my mind. It was because of that fertile imagination under God's guiding hand that I began to realize that all things were related, and that the universal nature of music in the human saga could not be explained simply as the result of human innovation and creativity and activity. It had to have deeper and more relevant roots than that, given the inherent near-impossibility of successfully defining our own human roots.

And then, there is this: the matter of God and His providence and His sovereignty and the over-arching existence of the Scriptures and other ancient sacred writings from the Judeo-Christian tradition whose verity really cannot be explained away. And what about the tomes of authoritative writings in the worlds of science and related disciplines, many of which make an all-out attempt to defy and deny those Scriptures. And what about the writings from other, even pagan, cultures which so closely parallel the themes and dogma of our Western culture? None of these can be dismissed unceremoniously as insignificant or irrelevant for the very reason that they so closely emulate the themes and dogma of our conservative Western beliefs.

The worlds of Narnia and Ea, as they flowed from the pens of C.S. Lewis and J.R.R. Tolkien, became favorite places of contemplation for Jana and me, because they were so vast and because they were the mysterious product of the thoughts and songs of the singing creator-gods of those worlds, so ingeniously visualized by Lewis and Tolkien. How could Aslan transform a thought into a beautiful song which could become green with bursting trees under a burgeoning blue sky?[11] How did the people of

---

11   C.S. Lewis, *The Magician's Nephew*: Copyright 1955 by C.S.Lewis, Ptc, Ltd.

Ea progress so easily from a thought hidden in the mind of Iluvitar and, then, simply expand into the intelligent, singing people of Ea and all the beauty of the earth around them?[12] [13]

There were other mysterious places in my mind—the world of music itself, for instance. I am grateful that I attended college late in life. By the time I reached forty-something, the usual confrontations of expanding intellectual acuity would not have the same potentially devastating effect that they may have had on a sixteen-year-old fresh out of high school. So I was more or less free to allow the process of learning, categorizing and formulating ideas of my own without the additional burden of trying to discern what was right and what was not right. Of course, I must make it clear at the outset that when I first began to put my ideas into a more or less coherent shape, I really did not have any well-defined direction for my project or even any clear idea what that project **actually** was. I only had the deep conviction that there was much more reality in the world of the spirit than we human beings could begin to understand and that reality had something—maybe everything—to do with God and the arts and most of all—with music.

And this is precisely where sanctified imagination and Jana took over. Her thoughts were always with me:

> "It makes no difference if you don't know exactly where you are going with a thought—you know they are all guided anyway, by something somewhere, and if you stay on the right track—and by the way, music and Scripture

---

12 It seems almost prophetic that Tolkien would express the idea that Iluvatar "made visible the song" in the opening pages of *The Silmarillion*.

13 J.R.R. Tolkien, *The Silmarillion*: Houghton Mifflin Company: 2001.

are good guidelines—you will not only find the way, but the reason for going that way. So dream on...."

The reader must not confuse my intimacy with Jana, the free-thinking, imaginative friend that was, in reality, my own fertile and creative mind at work, with fantasy. It is far from fantasy. It is a great gift to me from my God, Who gave it to me to use in framing and expressing ideas that may be, in the vernacular, "outside the box." From this point on in this book, when I speak of Jana, understand it as an invitation to every reader to engage the same inherent ability to think thoughts that may not necessarily be found in doctrine or in a seminary text. **It is an invitation to begin to glimpse possibilities**.

With these thoughts in mind, I believe it is appropriate to move on to the body of this book—*God's Opera*.

Act 1
LANGUAGE AND THE CHRONICLE TOLD

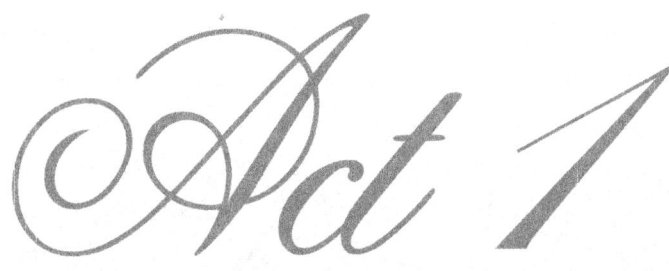

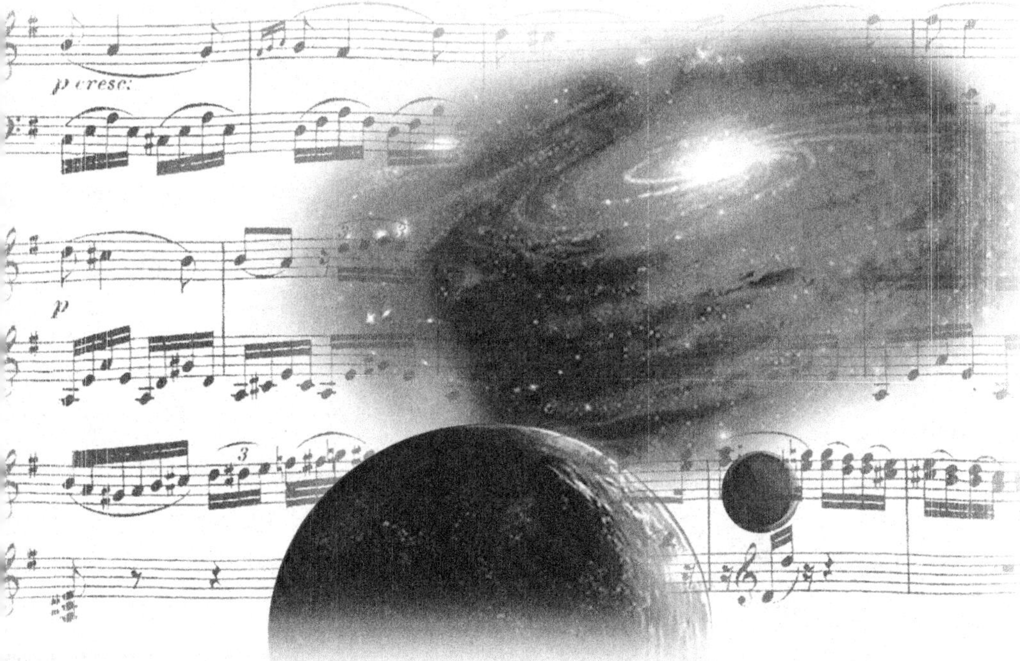

Scene I

THE OPERA—THE CHRONICLE BEGINS

> Consider this...
>
> *The unity of all of creation is astoundingly simple in its uniformity and balance.*

"*The preparation of an opera performance involves the work of many individuals whose total contributions sometimes spread across a century or more. The first, often unintentional, recruit is likely the writer of the original story. Then comes the librettist, who puts the story or play into a form—usually involving poetic verse—that is suitable for musical setting and singing. The composer then sets that libretto to music. Architects and acousticians will have designed an opera house suited or adaptable to performances that demand a sizable stage; a large backstage area to house the scenery; a "pit," or space (often below the level of the stage) to accommodate an orchestra; and seating for a reasonably large audience. ...The producer, conductor, and musical staff must work for long periods with the chorus, dancers, orchestra, and extras as well as the principal singers to prepare the performance—work that may last anywhere from a few days to many months.*"[1]

1   opera (2014). In *Encyclopaedia Britannica*. Retrieved from http://www.britannica.com/EBchecked/topic/429776/opera

God's Opera

THE LIE IN THE MAKING

As a grad student matriculated as a music "educator," I found myself to be out of sorts with the things I was being asked to take as the absolute basis for my teaching career. But I did not know why I was out of sorts with it all. It simply did not ring true at that time for some reason, which I would only discover slowly and much later, at that.

> "Music is organized sound. If a boom from any source is part of an organized sequence of sounds, then it qualifies as music…Music is a human activity created for people who perceive organized sound as interesting and psychologically valuable."[2]

This philosophy was part and parcel of the educational process in the 1970's. The enlightened, humanistic approach to education was at the helm of all theoretical development and experimentation at that time and has continued to reign as the high ruler of our intellectual and world views ever since. The entire premise, distilled to the bare bones, was that man was evermore progressing toward betterment and that human endeavor and achievement were the apex toward which we should all move. At the turn of the prior century, Friedrich Nietzsche had popularized his theories about God, education, music and theology and made the idea that "God is dead" an acceptable topic of conversation—and a feasible consideration, at that.

All of this pointed the path of human development in a very secular direction. The immediate effect on cultures all over the world was to set in motion a philosophy which would eventually erode and finally nullify the opposing philosophy which upheld God as Creator of all things—

2   Charles R. Hoffer: *Teaching Music in the Secondary Schools*: Wadsworth Publishing Company, Inc. Belmont, CA 94002: 1975.

including the standard of morality and ethics and the foundation of cultural absolutes which saw God as omnipotent and omniscient—and, therefore, the final adjudicator of all things. The overall consequence of this deviation from "God as Center" to "man as center" has completely altered the means of humankind's interaction with his entire universe, as well as his culture and his knowledge structures.

It is not difficult to see the effect that this secularization exerts against any idea of an Almighty God, especially One Who might be of a musical nature and might be the center of the entire creative process. Along with this burgeoning growth of secularization and what really amounted to worship of human intellect and worth, the nullifying and deadening effect of having no need for God since He is dead anyway, quite successfully shifted the paradigm from "God as the center of all that is" to the much more manageable paradigm of "man as the center of all that is."

James Burke, in his monumental book, *The Day the Universe Changed*, said this,

> "We are what we know—and as the body of knowledge changes—we also change."[3]

And so, we have become creatures of great pride and enormous delight in our own accomplishments and have virtually lost sight of the wonder of viewing our cosmos as the magnificent work of art that it really is—*God's Opera.*

In all probability, the "lie" should be defined at this point. How incredulous to go on with a theory, a reasoning over the beginnings of anything, if we do not identify the problem—the "lie." While it is not the

---

3   James Burke: *The Day the Universe Changed*: Little, Brown and Company: September 1986.

intent of this book, as I have said before, to present an apologetic for my own orientation toward conservative thinking, it remains imperative that we understand the logic which has directed my writing of this book to begin with. And so we proceed—

Some may bridle at calling the departure from God's perfect intent at the moment of creation—a "lie." For the sake of clarity, this is my own personal evaluation of the human dilemma in the 21st century—we are living in an atmosphere of human egoism fostered by this "lie." We exist in a world, here and in the cosmic arena, that believes that it has little, if any, need for God, a world that has a vague and ill-defined history of origins, a world filled with human accomplishment and self-styled grandeur. The concept of that world is also fraught with threats of uselessness because of the strength of the atheistic philosophy on which it is built. Eternity is an impossible commodity, because it cannot be measured or calculated or proven. Therefore, to think of an eternal God is to be soft-brained and unintellectual.

At the very best, some outside of the conservative, creationist community may allow for a theistic Being Who set things in motion in the beginning and then departed on a "forever vacation," never needing to be dealt with again. As this book progresses, it is my hope that it will become apparent, at least in the reader's "sanctified imagination," that this is not the whole truth or, more likely, it is not the truth at all.

How much more appealing is the imaginative proposal made by J.R.R. Tolkien and C.S. Lewis in their respectively magnificent tomes, *The Silmarillion* and *The Chronicles of Narnia*. The notion which both of these great writers held was that of a singing creator god whose creative thoughts, couched, in the beginning, only in their minds, grew until these thoughts became audible songs, and, when finally sung, these thoughts burgeoned into tangible beings and environments. Both Iluvatar and

Aslan, sang their cosmos into being and their creatures were the direct result of their singing.

It is also of great interest to note that the evil pride of the evil counter-forces in each saga created the songs that eventually polluted and, to whatever degree, ruined the beauty of the would-be perfection of Ea and Narnia.[4] [5]

It is my firm belief that these tales of fantastic beauty about creation and the creator gods were inspired by Tolkein's and Lewis' deep faith in the actual Creator God, in concert with God the Son and God the Holy Spirit, Who together comprise the Trinitarian God that we worship. This thinking is the warp and woof of my thinking as well. And—it is vital to remember the trinitarian construct which exemplifies both our musical constructs and the Trinitarian nature of our God. There is a resonance here which cannot be left unaddressed, and will be seriously considered later in this book.

THE ANTIDOTE TO THE LIE

What is vitally important is to keep in close touch with the purpose of this book. This is neither a manual on new theory nor an expose' to be seen as actual hypothesis or proven theory. It is only, as the psalmist, David, wrote in the 45th Psalm, a heart *"overflowing with a good theme"*—a recitation of *"my* (own) *composition concerning the King* (Creator),"* and filled with ideas for later scholars to take and flesh out, each one in his own discipline.

In addition, it is not the purpose of this book to present a theological apologetic of any kind. But, it is imperative to recognize that without a well-supported knowledge of the traditions and doctrines of our faith, the entire framework of these ideas would crumble. Whether or not one agrees

---

4   C.S. Lewis, *The Magician's Nephew*: Copyright 1955 by C.S.Lewis, Ptc, Ltd.

5   J.R.R. Tolkien, *The Silmarillion*: Houghton Mifflin Company: 2001.

with the premise that there was a profound shift of paradigm from **God as Center to man as center** following the dawn of the Age of Enlightenment, it is necessary to recognize that such a shift would, in any event, demand a rather profound change in the entire philosophy of life and learning in order to accommodate the change. Of course, this shift was gradual; not cataclysmic. It was simply an almost obscure ground swell of changing cultural sentiment viewing the preeminence of the Creator as mandatory and central, in favor of viewing the creature, mankind, as the driving force and the composer of all that was valid and important. Man's wits and intelligence afforded the pinnacle of achievement, as seen through the lens of secular philosophy.

There were apparently beneficial results from this change in attitude. Realizing the potential of human intellect undoubtedly fostered the growth of all of the disciplines of learning, the sciences of discovery and the establishment of the means for almost endless accomplishment. This cannot be seen as detrimental except for the fact that the further down the road of human success we travelled, the further away we strayed from the comprehension of God as Creator and Designer **and** Master to the position of seeing our own accomplishments as grand enough to qualify humanity as central to the entire understanding of our universe and its inexplicable complexity.

This notion has fostered the veiled philosophy that mankind is, in fact, always improving, growing, moving forward to a more and more pure understanding which will eventually give him the entrance into holiness and perfection that he is sure will give him, in addition, entrance into whatever vague idea he may have of being all-powerful—or extinct.

Initially, the shift was not visibly overpowering. But, by the 21st century, that shift of paradigm has resulted in dramatic changes in our understanding of nationalism, morality, economic ethics, medicine, family and many other peripheral issues and has given man the sense that he has the right to control these things if they impinge on the most imminent issues which he holds more important at any given time.

THE OPERA—WHY

There is no other art form as all-encompassing as opera. It often goes by the pseudonym, "grand opera," precisely because of its bigger-than-life dimensions, obvious from the subject matter of the story lines to the hugeness of the cast and the opulence of the sets and expansiveness of the music. To be sure, early opera did not command such a reputation, but the very nature of operatic expression did demand that the structure of the opera would grow into its eventual shape of opulence and grandeur. All artistic performances require minute attention and detailed planning in order to bring as much accuracy and organization of beauty into that performance as possible, but no other that I know of can equate to the demands of the operatic feat.

In fact, it is probably the very intensity of this particular art form that may have been responsible for its eventual demise in the minds of many average folks. But I am going to challenge the reader to lay aside that prejudice, if one is indeed present, in the service of being able to see this thing—opera—as the unmatched and incredibly expressive and beautiful musical enactment that it is. I was discussing this idea with a friend who was not sure that the opera would make a good "foil" for the analogy to creation. One reason for her trepidation was the "fat soprano." Initially, I laughed, but my friend was adamantly determined to tell me why she felt that way. She had good reason:

> "The whole idea of the diva—the prima-donna—and the larger than life sound of the lead female character (which

the soprano usually is!) runs counter to what we commonly think about when we think about creation and God. The very sound of the diva soprano voice is self-centered and arrogant, I think. Maybe you should find another "foil" to reflect your musings."

No other "foil" would fit what I was beginning to see as the vast, inexplicable wonder that is, was and continues to be creation and the immensity of the universe(s) that are the result and supporting vehicle for that creation.

Wagner may have sensed this dichotomy in aesthetics when he began to write the aria out of his huge operatic creations. He needed for the epic itself to be primary—and not the lead soprano. So, not only did he eliminate the soprano aria, but he dropped the aria almost entirely as a necessary component in the opera and substituted what became known as the "leitmotif." Simply put, this is a short, melodic fragment, which he relates to a particular character or event in the total musical tapestry of the work. Actually, this new technique did not serve to make the opera less opulent and lavish. By contrast, it only made the musical fabric more thick and the whole event more complex.

Wagner also began to employ the use of the "schena,"[6] which amounts to an elongated series of scenes and aria-like solo passages, which directly contribute to the story-line and the depiction of the action at hand and become far more integral to the storyline than the solo aria.

I was afforded the great opportunity of seeing opera at work firsthand in Spoleto, Italy, when the great Italian director/producer, Visconti,

---

6 SCENA (SCHAY-nah): Literally "a scene;" a dramatic episode which consists of a variety of numbers with a common theme. A typical scena might consist of a recitative, a cavatina and a cabaletta. An example is the "Mad Scene" from *Lucia di Lammermoor*. Info@operaamerica.org: All content Copyright © 1995-2014, OPERA America, LLC.

directed *Manon Lescaut* by Puccini, there at the Festival of Two Worlds. The Westminster Choir served as the choir-in-residence at the festival and the center-piece of our performances was as the chorus in that production, which was mounted for two consecutive summers. The opulence of every emotion, every gesture, every note sung was all but overwhelming.

The costumes were each a work of art. I remember that the abbotti, the abbotts or religious men, who came to visit Manon and her aged lover were dressed in rustling silk habits of intense pastel blues and greens. Minutely pleated cuffs protruded from the sleeves of our frock jackets and lavish gold buttons cinched the jackets closely to our bodies so that we could only move with great precision so as not to fall flat on the stage. Huge gold buckles on our shoes made walking a real experience. Yes, ladies did play the part of some of the men, maybe to better emphasize the asexual nature of such personages.

The point is that the entire production was so sumptuous as to be unabashedly overstated. But, you see, that is the very point, and it is hard to make the comparison. Opera intends to be larger than life, because it sees its own being as just that—larger than life. Our universe and what-ever else may exist inside and outside of it is so unabashedly glorious, huge in all its dimensions and so incomparable to anything else we know, that it takes some vastly grand and unabashed analogy to even approach being able to express that glory. It is incumbent upon us as we try to apprehend and understand the magnificence of creation that we do not, in any way, demean the beauty and holiness of the creation we seek to understand. It is only important to look at the reflection of that beauty in another venue in order to catch a glimpse of its magnitude and the opera seems to give such an opportunity because of its own hugeness.

This points the way to another thought, which is inherent to the reason for even trying to make the comparison. Long ago, it occurred to me that "nothing is for nothing," as the song goes. God had good intention for every single minute creation for which He is responsible. By the end of this book, I hope it will be evident in many different ways and by many different means that God, in His wisdom, saw to it that we human beings have an abundance of opportunities to compare the truths of one segment of life to the truths of another segment. Everything is, after all, related, and Einstein knew that all too well. That is what drove him for a lifetime to seek after the "theory of one" that would explain the all-encompassing wonder of all that is. Of course, he never found it. But he did give mankind an enormous legacy of knowledge upon which to build.

THE OPERA—HOW

The *Encyclopedia Britannica* article on opera, its structure and developmental necessities makes it quite clear what massive forces are called into play in order to bring an opera finally to fruition. The list is long and varied. It is precisely here—as we view this list—that the analogy of opera to creation or creation to opera begins to break down. No analogy known to man can be faultless. This analogy is far from perfect, but it works as long as one keeps his "sanctified imagination" intact.

The reality of creation as an eternal, holy event presupposes God as original writer, librettist, composer, architect, acoustician, producer and whatever other creative personages are necessary to the event. Mankind enters the stage only by God's command or intent and from the point where man is himself created, he may be seen as the musicians are seen as God equips them. However, they do become integral and inherent elements in the performance of the "opera."

The genius of Tolkien[7] and C.S. Lewis[8] was that they could see the creation of Ea and Narnia as musical phenomena which had their seminal beginnings as ideas in the minds of the singing creator-gods. Those ideas were alive only there—in the minds of Iluvitar and Aslan—they were actually non-entities until they were **thought**. The thoughts became songs; the songs became the realities of whatever sort the creator-gods would imagine—beautiful surroundings, clouds, earth, water and other singing beings. In these fantasmic places, Ea and Narnia, each new singing being added more to the environment of beauty.

There is one item of utmost importance: In each imaginary world, very soon, an evil god appeared, and that evil god swore to make more beautiful music than either Iluvitar or Aslan could make. That evil music sung into the exquisite beauty of each surrounding creation immediately spoiled the beauty and started a cycle of increasing corruption. Interesting, in the light of what we know about sin and Satan.

This brings up a question which will be addressed in the ACT on the theology behind the theories presented here: both the books of Deuteronomy and Job illude to the presence of Lucifer as an Angel of Light who was present in the heavenlies before time was. This issue stands as more of a theological/philosophical issue, and it is not the intent of this writing to provide an apologetic for such issues. This is precisely the place where the "sancitified imagination" can make an intelligent excursion into the worlds of "maybe"—but not "make-believe."

One could say with reasonable confidence, that once God the Creator

---

7   J.R.R. Tolkien, *The Silmarillion*: Houghton Mifflin Company: 2001.

8   C.S. Lewis, *The Magician's Nephew*: Copyright c1955 by C.S.Lewis, Ptc, Ltd.

had completed the original six days of work and seventh day of rest,[9] everything that was needed for life and knowledge to flourish had been put into place and from there, the creatures of God, mankind—made in the image of the Triune God—were charged to: *"Be fruitful and multiply; fill the earth and subdue it; have dominion over (it)."*[10]

Basic to the mounting of any opera is the story line, which usually is quite complex and may even necessitate a secondary or even third plot. And just so is the chronicle of the creation of the universes. This is a tale not told in short paragraphs with words of one or two syllables. The secular world holds to an extremely convoluted, if not labyrinthine, theory of evolving cosmic and organic development which, because of the very nature of the theory, demands a timeline of millions, if not billions, of years. Again it is not my intention to argue the point, or even to try to explain why my theories differ from the post-modern view of creation. At any rate, the story line of the opera and the story line of creation have many similarities—and many defining differences. It is my belief that the original writer and the librettist of creation are One, as opposed to the common concept of many artists and artisans co-operating to create the sweeping saga of an opera. Nevertheless, the successful writer and librettist must be able to see the "big picture"—he must have a clear view of the entire scope of the story in order to keep the lines of each plot tightly woven and must never allow himself to wander from the grand and concise epic that he is portraying so that the end product maintains its clarity

---

9   There is considerable difference in theological preferences concerning the length of the creation saga. I remind the reader that this writing is in no way attempting to address these issues definitively.

10   All of the necessary systems, laws, elements, and energies were immediately in place at the end of God's seven days of involvement with the creative process. Man does not—as I have written before—invent or create anything. We are only equipped to discover, emulate and, in some sense perhaps, co-create with the tools that Almighty God has gifted to us.

and is easily appropriated for the needs of the next artisan. At precisely this point, I am compelled to make this observation: at the grand moment of initial energy output which I hope to show as God's initial split instant of unimaginable creative power and song, He had already conceived of, planned and articulated every concept, every law, every sequence which would become reality at that moment and would continue to develop and sustain all of creation for as long as God Almighty deigned to support it before putting an end to what we call temporality—TIME. In the mind of Creator God, everything was complete at its conception.

In all probability, the next artisan would be the composer of the music in the temporal world. Without any further consideration, we can immediately see that, in the matter of the creation of the cosmos, this artisan is also in the Person of Almighty God. The vastness and beauty of such an all-encompassing epic as the creation could only be accomplished by the Omnipotent and Omniscient God, if one takes any of the Holy Scriptural Writings seriously. We have very quickly arrived at a place of impasse— one can never draw the comparison of eternal to temporal. In fact, that is one of the cardinal aspects of this world's sad departure from its root "first cause."[11] So, we must quickly abandon the comparison between opera and creation and begin to allow our "sanctified imagination" to open the window into seeing the comparisons which we need to see and those that are not decisively important.

---

11 First cause, in philosophy, the self-created being (i.e., God) to which every chain of causes must ultimately go back. The term was used by Greek thinkers and became an underlying assumption in the Judeo-Christian tradition. Many philosophers and theologians in this tradition have formulated an argument for the existence of God by claiming that the world that man observes with his senses must have been brought into being by God as the first cause.

Encyclopaedia Britannica Online. 2014. Web. 12 Sep. 2014.

http://www.britannica.com/EBchecked/topic/208087/first-cause.

Further on, we will be able to see mankind in the roles of the architects who build the opera house, the acousticians who design the stage and hall itself, the musicians, singers, dancers, conductors and rehearsal technicians who give temporal life to the viewing of the epic itself. But all of this endeavor will require constant care over the "sanctified imagination" that gives opportunity to see and to understand this incredibly extra-ordinary, ongoing, all-encompassing recital of God's unattainable creative energy and power.

Considering creation as the most magnificent and grandiose opera of all time is almost trite and the analogy has broken down very quickly. But the analogy remains the "perfect foil"[12] for setting the stage for the opera as a way to talk about and communicate about the vastness and magnificence of creation without becoming involved in semantic discussions, especially where the matters of evolution and religious predilection are concerned.

The post-modern world is very much taken with the idea of the "big bang," even though no one has been particularly successful in describing what that event was and how and why it happened. It just "happened." And from that moment on, the evolution of "man from matter" began along with the supposed evolution of all of our present environment and its immeasurable intricacies from the "primordial slime." Again, this writing is not geared to addressing any of these theories in a critical way. It is a matter of setting boundaries for understanding.

THE GREATEST OPERATIC PERFORMANCE OF ALL TIME

Here is one of my favorite speculations: Science has now made the all-but-certain observation that if there were no sound, there would be no

---

12 Ibid. *God's Opera*: Overture pg x.

matter. It is sound that holds all matter together and, in fact, matter itself is actually made up of nothing more nor less than "strings" of vibrating energy, according to the string theorists.[13]

And the emerging science of cymatics is presently making the claim that without sound, matter would not even be able to exist.[14]

My speculation is this: If it is true that matter is comprised of nothing but "strings" of vibrating energy; if it is true that it is sound (vibrating energy) which holds all matter together; if we can then conclude hypothetically that it is sound that is the raw material of all creation, then where did this sound come from? The entire consideration of "first cause," the eternality of God and the Scripture that describes God's speaking, *"...and let there be...,"* and the ensuing words, *"...and it was so,"* takes on a entirely new patina of possibility. It becomes infinitely more possible to ask the question, "Is the genesis of this sound, this vibrating energy, this new vision of the raw materials of creation actually directly from the singing of God Almighty?" And if this is even a

---

13 Think of a guitar string that has been tuned by stretching the string under tension across the guitar. Depending on how the string is plucked and how much tension is in the string, different musical notes will be created by the string. These musical notes could be said to be excitation modes of that guitar string under tension. In a similar manner, in string theory, the elementary particles we observe in particle accelerators could be thought of as the "musical notes" or excitation modes of elementary strings.

In string theory, as in guitar playing, the string must be stretched under tension in order to become excited. However, the strings in string theory are floating in spacetime, they aren't tied down to a guitar. Nonetheless, they have tension. The string tension in string theory is denoted by the quantity $1/(2 p a')$, where a' is pronounced "alpha prime" and is equal to the square of the string length scale. The Official String Theory Website: http://superstringtheory.com/basics/.

14 Vibration underpins all matter in the universe. No matter can exist without sound and vibration. To see the periodic motions that lie at the heart of matter is to lift the veils that conceal many mysteries of the universe. The CymaScope represents the first scientific instrument that can give us a visual image of sound and vibration—a cymatic image—helping us to understand our world and universe in ways previously hidden from view. Copyright 2008-2013 cymascope.com. All Rights Reserved.

remote possibility, to be examined in the privacy of our sanctified imaginations, how are we supposed to approach it; and, further, how can we analyze it and eventually see it shaped into an acceptable thesis—to be eventually proven? Scripture and things of faith cannot be empirically proven, can they?

As a trained singer, I am fully aware of all of the mechanics necessary to producing a beautiful singing sound, and I am **fully** aware of the energy intrinsic to the making of that sound. Christendom and Judaism have long been familiar with the idea that the creation saga is based on God "saying" something of a creative nature which put into action something mystic— something that resulted in the appearance of light, dark, water, land, animals—and people. Spoken sounds have the same geneses and the same need for physiological structures in order to produce them that sung sounds have—but to a much less demanding and energetic level. And it is interesting that the Hebrew language allows for the parallel translations of "word" and "song" in the Hebrew word *dabar*.

What if creation **is** the result of a grand "Oper`a Dei"—*God's Opera*— sung by Almighty God? Giving a voice to this theory is the most important reason for this book.

But there is one more mandatory caveat: Opera as an art form is a more or less linear experience. The story line begins, unfolds and ends. There are depths and heights of musical and dramatic intensity, but the story line continues and finally comes to some point of climax or complete fruition. The tragedy or the comedy is consummated, and with a great flourish, all is well. Not so with the unfolding of this new concept which is the idea of the whole magnificent and all-encompassing review of the cascade of entwined events which is the creation saga.

Act 1 — Scene 1

It is far more related to the concepts of Pythagoras and his grasp of the unity of all of creation as seen through the understanding of the Golden Ratio. This idea of the constancy of design and the relatedness of design and the inviolability of design in all of the universe's myriad expressions can only be understood as compared to the unfolding of the fiddlehead fern, for instance. Convoluted, complex, intricate—and yet astoundingly simple in its uniformity and balance. Proportional harmonies in nature.[15]

---

15   *The Power of Limits*: Gyorgy Doczi: 1981: Shambala Publications, Inc. 300 Massachusetts Avenue, Boston MA 02115.

Scene II

WORDS—WHAT ARE THEY?

> Consider this...
>
> *Think of the scope of possibilities if one could apprehend the immeasurable link between words/songs and energy!*

"*—every man is confronted by the Word of God, spoken directly to him in his own concrete situation, and demanding a response in terms of decision—a life-and-death decision for or against God himself. As Martin Buber has put it so vividly, God's question, 'Adam, where art thou?' is addressed anew and inescapably to every human being—When the religion of the Bible is thus transposed back into its original form, that of a dynamic and dramatic encounter between an initiating God and reluctant men who try to flee from his presence, then the attempt to reduce Christianity to a set of rational propositions, and to reduce God to 'Being-itself,' actually appears to represent one of man's more sophisticated efforts to evade this encounter. For to render Biblical religion in such an impersonal (and therefore subpersonal) way is effectively to neutralize it. It becomes something which I can hold at arm's length, analyze, and evaluate. I can take it or leave it. And, conversely, the one thing I cannot do vis-a-vis the Word directed to me personally is to sidestep it. Confronted with the summons, 'Choose ye this day whom ye will serve,' I may not remain on neutral ground. Any attempt to ignore the alternatives constitutes by default a decision against God.*"[16]

16 http://www.philosophy-religion.org/cherbonnier/theology.htm: Edmond La B. Cherbonnier.

WHY WORDS?

s I considered the recounting of the chronicle of creation, two questions came to my mind immediately. First of all, is it necessary to establish an apologetic for the Scriptures, for the Word of God? The definitive answer is "no." It is not the intent of this book, as I have said before, to defend the veracity of Holy Writ. But it is necessary to make it quite clear that the theology of the conservative church, explicitly understood is, in fact, the original source of my theory of creation as a musical event. The problem may be that this explicit understanding is so often overlooked.

Secondly, is it necessary to trace the origin of words back to the beginning of all things as we know them? I believe that the only answer is "yes." Re-read, and ponder, with the vitality of your "sanctified imagination," these words from Chebonnier's paper, The Theology of the Word of God:

> "When the religion of the Bible is thus transposed back into its original form, that of a dynamic and dramatic encounter between an initiating God and reluctant men who try to flee from His presence—the attempt—appears to represent one of man's more sophisticated efforts to evade this encounter."

THE ORIGIN OF WORDS

Some of us may be shy about consorting with our imaginations as if they were living entities, or may avoid considering an alter ego as a means of being better able to mirror our thoughts in such a way as to make those thoughts more lucid. I am not. And if you would recall my allusion to

Act 1 — Scene II

Jana, who has been a constant part of my journey in penning thoughts on creation as a musical event of cosmic proportions, you will realize that she actually has enabled me to align what I wanted to write and to winnow out what I did not want to include in these writings. Sometimes, the best companion is one's most inner self, and if you must give that personage a name, then by all means, do so. Remember that this is a journey into the land of "maybe"—not the land of "make-believe."

Imagine suddenly coming to life in the Garden of Eden. Perfect beauty, perfect climate, perfect light—all things perfect. Then you suddenly hear a powerful, commanding voice that seems to fill the air everywhere! The phrase, **"an initiating God"** is evocative and effectual.[17] It suggests a God Who thinks, cares, communicates and personifies all things beautiful and orderly.

You have no recollection of **anything** before hearing this voice—primarily because you WERE NOT before this voice caught your attention! And it caught your attention only **after** the Creator God had given you life, by transmitting His living breath into you. How did this take place?

The ancient Scripture says: *"Then God said, 'Let us make man in our image, after our likeness; and let them have dominion—So God created man in His own image—male and female created He them. And God blessed them, and God said to them,' Be fruitful and multiply, and fill the earth and subdue it; and have dominion (over it)."*[18] The inference is, of course, that the creatures of this **"initiating God"** would be as perfect and beautiful as He Himself inherently **is**.

---

17 Colossians 1:16-17: Revised Standard Version: *"For by Him all things were created—and in Him all things consist."*

18 Genesis 1:26-28: Revised Standard Version: Excerpted.

If only mankind could, or would, spend as much time **pondering the possibilities** of an eternal God lovingly hovering over His brand new creatures and forming within them all that they would ever need to live a sumptuously bountiful life in a sumptuously appointed garden as mankind spends in **formulating so many other theories which cannot be proven either**, what do you suppose we could discover about ourselves?

The ramifications of such unfettered, sanctified imaginings are limitless. First of all, what was this voice like? Did it sing rhythmically? Did it change pitch? Was it frightening? Did it use words, and if it did, what words were they? What language? There is deliberation in some intellectually linguistic communities, that the language of God is, in fact, ancient Hebrew. And there is evidence to substantiate that idea. And, as a matter of fact, if you stop to think about it, words and communication had to have some kind of beginning somewhere. And in the realm of "first cause," that beginning was not bound by the intelligence of mankind. I am convinced that mankind is a **great discoverer** of beauty, science, art—all things—he is not the creator. So, when mankind in the persons of Adam and Eve heard that glorious voice for the first time, what was it like?

"The secret things belong to the Lord our God, but those things which are revealed belong to us and to our children forever."[19]

Scholars are very clear that God's Word, in whatever form, is imparted by revelation and may be imparted in whatever way is divinely discerned—dreams, visions, by angelic beings—or spoken word. Understanding the necessity and legitimacy of communications from God is vital to the rest of the consideration of creation as a dynamic and all-powerful event. By that I mean that whatever shape or construct divine communication eventually or differentially takes, it is to be

---

19 Deuteronomy 29:29: New King James Version.

understood as being defined by **endless energy**—an energy which we of humankind cannot by any means apprehend.

When I was a youngster in school, the character of energy was a constant enigma which I could not wrap my sensibilities around. What did Mr. Peck have in mind when he said in chemistry class that energy could not be destroyed and could not be created?—It could only be transmuted from one form to another!

If the reader can allow his or her sanctified imagination to extend to that place of possibility—and allow his or her conversational alter ego into dialogue, it will not be too foreign a thought to realize that this energy could have come into being as part of the idea of "first cause"—that the never-created God, Who always **was**, formed all things at once in that mind-cracking instant when the first sound emanated from His being. And that the remainder of all things coming into being was simply a result of that one, omniscient, all-encompassing, all-inclusive moment of eternal energy being released at the will of God and God alone. Imagine an **energy** that all-encompassing. And imagine the energy that He created for our use and good—entirely complete at that moment of creation.

It is my belief, after over forty years of study, prayer, thought and consultation, that this energy was in the form of, not-only word, but also song—or just maybe, entirely song. This theme will be pursued all through this writing.

THE TRUTH ABOUT WORDS?

Ancient Biblical Hebrew may be viewed as a living language, in that, much of its meaning depends upon context. Emphasis in this language is only conveyable by repetition since there are no comparative adjectives in its grammatical structure. Without going into a lengthy explanation,

simply put, the creative and fluid possibilities of the language are limit-less. Of, course, that does not make for an easy task if you are the trans-lator. But it does allow for exciting consideration of how words, or, as we shall see later, songs, figure into the fabric of the theory of a Creator God Who created by the expressive beauty and power of His thoughts, which became His Word—or His song.

And, it is also important to keep in mind that what I am extending to my readers, and hopefully, to other scholars better equipped to follow a particular thread of thought than I, is that these nuances of thought can be explored, and can provide paths to be followed, theories to be shaped. And, consider that this exploration may deepen our understanding of the power of words and the extent to which they may reach in the process of life itself. **No accident** in the existence of words or their development.

Another theme which is far more difficult for me to define is the theme of "relatedness." When I say that "all things are related," a variety of interpretations surface immediately. And depending upon the passion of the scholar for his own field of knowledge, there is room for one to say, "Nonsense! Quantum mechanics can have nothing to do with the beauty of a Bach fugue, and neither one has anything to do with the literary scope of a Shakespeare play." On the surface, that may be true, but by the end of this book, it is my sincere anticipation that the "relatedness" which I see in all things will become an acceptable concept, a concept based on the apprehension of the idea of "first cause" and the unifying effect of the idea of a musical foundation under-pinning all of creation and thereby providing a possible basis for the idea of Einstein's "theory of one—one simple, beautiful set of laws which would govern the entire universe."[20]

---

20 *The Elegant Universe*: Brian Greene: W.W Norton & Company Inc. 1999:Preface, pg. xi.

Act 1 — Scene II

It is at this precise point that I choose to explore the idea of "words" and their genesis and ultimate meaning in the overall scheme of our cosmos. They—"words"—have become the changeless and inexpressible fabric of all communication and learning. Without words, it would be a strange world, indeed. Even the uneducated must eventually come to grips with "words" in order to communicate. The history of words is as much a part of the theory of creation set forth in this book as is the idea of musical raw materials being the shape and fabric of all matter. It may seem inconsequential, but even the idea of spoken words has a musical component. It is a dull public speaker, indeed, who allows himself to be a "monotone."

Again, we call upon the sanctified imagination and engaging our inner self, alter ego, in order to get in touch with a deeper, more elegant appreciation of what our languages contribute to our lives and how that may have come about. The first conversations between God Almighty and His two favored creatures, Adam and Eve,[21] was possibly the most beautifully simple conversation ever enjoyed by anyone. It is probable that God had so much to share with these two new beings, that it took quite a long time to set things into some sort of context. Then, there would be the desire to explain and describe all the things that these two were beholding. And, there were the instructions which had to be given, as well. But, the language was clear, uncomplicated, straight-forward **and**, I wish to believe, very beautiful, lilting and aurally pleasing, whether it was a long, detailed dialogue or a more immediate reciprocal matter of understanding one another. Obviously, God had to be the Initiator, Inventor and Interpreter of all of it. Did He have to teach them how to speak? Did He have to school them in grammar? There is no end to what we could imagine.

---

21 I take a deliberate, literal view of the Book of Genesis as an historical account of early events in the creation and development of our cosmos. But again, I say, it is not the liability of this writing to provide an apologetic for my theological views.

But I do not think it is all that important. What is important is that the first language, as the creation itself, emanated from the Creator God and was His unique language for interacting with His creatures.

Time went by and people lived and died and farms were planted and music was sung and metal-workers smelted and built, and the effect of original sin put its stamp on the cosmos. Understand that there are theological questions of all kinds which need to be recognized as valid questions but are not the focus of this book and from this point on will command less and less attention in deference to the idea of music itself in creation. Suffice it to say, the words of Holy Writ, the Scriptures, especially those dealing with the process of creation, are the very fabric of my personal belief system concerning that creation.

By the time of Genesis 11, even the Noahic flood had taken place and earth was supporting generations of a burgeoning population.

> *"Now the whole earth had one language and one speech—And they (the sons of men) said, 'Come, let us build ourselves a city and a tower, whose top is in the heavens; let us make a name for ourselves—But the Lord came down to see the city and the tower, which the sons of men had built—And the Lord said, 'The people are one and they all have one language—Now nothing they propose to do will be withheld from them—Let us go down there and confuse their language.' "*[22]

---

22 Genesis 11: Man has a propensity for self-seeking and an appetite for power. The essence of this story is that in response to this aspect of mankind's make-up, God decided to confound their speech so that they could no longer understand each other, which would, of course, limit their power against their Creator. The necessity here is not to prove or disprove that, but to provide a narrative for the development of various languages. Another opportunity for discoursing with your own imagination, if the literal interpretation of the Bible is not compatible for you. And how interesting that it is inherent in the power of words that all learning, good and evil, comes to pass.

Act 1 — Scene II

It would become burdensome to try to trace the evolution of communicative skills. Too much time has elapsed, too many variants have morphed into too many actual languages and dialects, too many nationalities have emerged and taken their legitimate place in the sequence of things. What is vitally important, though, is to bring into focus the emergence of words as communicative chattels and what part they play, when recognized, in the energy exchange which is perpetual in creation. As I have said before and will no doubt repeat, my personal starting point is distilled in the idea of "first cause"—the **non-created** God—the All-knowing and Omnipotent God, Who was, is and always shall be, and Who existed in all of His glory before the worlds were. The omnipresence of energy in our cosmic complex **may be** an effective allegory for His Omnipresence, as well. And the birth and growth of **words** may be an interesting aspect of that allegory.

Jana, my innermost well-spring of intellectual considerations, can be counted upon to ask questions that I, as a fairly intelligent, but conventional thinker, may miss entirely:

> "What does the immutability of energy have to do with the development of words and spoken communication?"

> Not that Jana would think it to be irrelevant or unrealistic, but simply that this would be a consideration that, on the face of things and for most thinking people, **has little relevance**. The connection between etymology and the transfer of energy seem, initially, to have no connection whatever. But the "Jana" in me had to ask the question and cannot be satisfied with a platitudinous answer.

The nugget-seed of curiosity has been planted in my brain and now I can't let it go. As a trained singer, I have understood the dynamics of

29

vocal sound for a very long time. The only parallel that I could draw for many years was that both speech and singing emanated from the same vocal mechanism, the larynx—the voice box and vocal chords, activated by the flow of air from the lungs, but these two functions were, of course, seen as just that—two separate functions and were, therefore dynamically different—or, were they?

Two ideas, from two differing academic disciplines, over a prolonged period of time finally succeeded in moderating my understanding:

- The study of Biblical Hebrew, where I learned that several Hebrew verbs had the same concomitant meanings of "to speak" and "to sing" and could be used interchangeably depending upon the context. The ramifications of that bit of knowledge are complex and will be dealt with in a later chapter. Nonetheless, this was a stunning realization. And, it led to a better understanding of the dynamism of "song" as opposed to simple speech in the creation process.

- The emergence of a new scientific endeavor centered in the UK involving the study of sound as a healing medium and the relevance of mathematics in the discovery and comprehension of sound as a cohesive element in all of creation. This new discipline is called cymatics,[23] a word derived from the Greek word, meaning "wave" or "billow."

23 Cymatics (from Greek: κῦμα "wave") is the study of visible sound co-vibration, a subset of *modal phenomena*. Typically the surface of a plate, diaphragm, or membrane is vibrated, and regions of maximum and minimum displacement are made visible in a thin coating of particles, paste, or liquid. [1] Different patterns emerge in the excitatory medium depending on the geometry of the plate and the driving frequency. The generic term for this field of science is the study of modal phenomena, retitled Cymatics by *Hans Jenny*, a Swiss medical doctor and a pioneer in this field. The word cymatics derives from the Greek 'kyma' meaning 'billow' or 'wave,' to describe the periodic effects that sound and vibration have on matter. http://en.wikipedia.org/wiki/Cymatics.

Recently, the cymaticists have made this provocative assertion:

> "Vibration underpins all matter in the universe. No matter can exist without sound and vibration. To see the periodic motions that lie at the heart of matter is to lift the veils that conceal many mysteries of the universe. The CymaScope represents the first scientific instrument that can give us a visual image of sound and vibration—a cymatic image—helping us to understand our world and universe in ways previously hidden from view."[24]

It takes a fairly high degree of imaginative insight to connect some of the ideas which I am presenting with the help of Jana, my alter ego, and her incisively inquisitive thinking. Such flights of visionary fantasy could well be misunderstood in the work-a-day world, if it were not for the existence of imagination in all of us. The trouble is that so many of us are unwilling to allow that God-given gift to flourish, and that is where Jana, as you know, entered the picture as the one with the great freedom to envision things "beyond" and, yet, keep that imagery closely monitored by spiritual candor.

Before we progress any further, it is important to connect a few thoughts which may not be visibly related. We understand the earliest origins of vocal communication to some small degree, at least. It is possible to comprehend how the Creator God might have formulated the need for the process of and the culminating effect of vocal exchange of ideas and thoughts. All of this, which may be slightly difficult to apprehend and to

---

24 Copyright 2008-2014 cymascope.com All Rights Reserved. http://www.cymascope.com/aboutus.htm.

be seen as an obligatory thread in the sweeping fabric of creation et al, was spontaneously and completely put into place at that incomprehensible moment when God spoke or sang the first syllable and all things **became**.

Again, allow me to call attention to the concept of energy expenditure in the act of singing. To fill an opera house with glorious sound without the aid of amplification, which, of course, was the norm for hundreds of years, the singer must first envision filling his/her entire being with the life-and-sound-giving air which will **energize** the vocal mechanism and excite all of the air-particles surrounding the singer in order to produce a beautiful and dynamic sound which some philosophic beings contend will go on forever. If one could allow oneself the luxury of sanctified imagination capable of transmuting that sound out of the temporal and into the eternal and sacred—the voice of God, not man—one could envision the birth of an energy incapable of human deciphering.

What then? Is the inception of energy bound up in the voice of God, the sound of His song, the power of spoken syllables shaped into words? Think of the scope of possibilities if one could apprehend the immeasurable link between words/songs and energy! Or words/songs and power! Or words/songs and creation!

Has Jana successfully inspired more deeply investigating the truth in which words are steeped?

Scene III

LANGUAGE—THE METAMORPHOSIS

> *Consider this...*
>
> ## Can it really be that a mysterious, unexplored language holds secrets that until now have only been known by the scholars of ancient Jewish lore?

From the time that I could listen and reason, I found that I had a restless mind that could not satisfy itself with the ordinary, the typical, the mundane. That is the very reason that Jana and her all-encompassing inquisitiveness became my cherished companion in my own sanctified imagination early on. She gave me a platform on which to examine things that I may otherwise have been discouraged from examining. But no one could take my own private thoughts away from me and I was free to examine them with Jana. And so I did. I could allow myself the luxury and freedom of asking questions that may have otherwise been off limits to the fundamentalists of my culture and my faith structure, in my southern home. To have a mind that is free to roam and consider new things, even in the face of strict cultural and traditional mores, is a great gift. This chapter may wander into places heretofore uncharted for some. But consider some of the most beautiful treasures that we have in nature. The process by which they came into being is often long and maybe even painful, but never unworthy of the journey from conception to fruition. One only has to be willing to take the time and energy to allow perceptions to change and to give space to new

thought. It is my sincere desire that by the end of this chapter many new ideas may have surfaced and that they may allow the readers a glimpse into the exciting journey I have been on for so long and thereby be inspired to explore newly discovered secrets of their own with God's confirmation.[25]

THE COCOON IS SPUN

"*And they said, 'Come let us build ourselves a city—let us make a 'name for ourselves,' lest we be scattered abroad over the face of the whole earth—But the Lord came down to see the city and the Lord said, 'Indeed, the people are one and they all have one language, and this is what they begin to do; now nothing that they propose to do will be withheld from them. Come, let Us go down and there confuse their language, that they may not understand one another's speech. So the Lord scattered them abroad—over all the face of the earth, and they ceased to build the city. Therefore, its name is called Babel, because there the Lord confused the language of all the earth—and scattered them over the face of the earth.*" Genesis 11:4-9

When a caterpillar spins its cocoon, I wonder if it has any hint of how beautiful it will become one day? And what its ultimate gift to the temporal world will be? Does it know that when that chrysalis finally cracks open, the result will be an incredibly crafted wonder of nature—a very delicately-winged flying creature who has the inherent ability to navigate many miles to a place of warmth and splendor to initiate the mating and creative season of another generation of creatures just like her?

Consider the inauspicious beginnings of this extraordinary creature. Hundreds of sibling eggs laid on the underside of a milkweed leaf or in

---

25 My thoughts about the journey may be just as important as the journey itself, but they are not necessarily a part of that chronicle. And it may help to think of this as a journey when you realize that things do change and seeing and voicing acknowledgement of those changes can then become integral to the adventure.

the grassy canyons of a field—**somewhere** specially chosen by the mother butterfly to provide good food for her offspring, because by the time they are able to feed on their own, she will be gone.

One of the themes running through the chronicle of creation is that nothing is random or coincidental. In fact, the idea of coincidence has faded into the far distance of perception for me. God's plan was, is and always shall be so perfectly designed that "nothing is for nothing"—that is to say, when He gives us the beauty of the butterfly, for instance, it is not an isolated event. Many parallels can be drawn. Many lessons may be lying just under the surface to give us new and remarkable insights. Many examples of our own inherent beauty may be dimly veiled allowing ourselves just to look and to imagine.

I like to think that if the tiny larva nestled inside that egg hanging under the milkweed leaf could look ahead to the magnificent flying creature it would become, it would be hard to believe. "Simply impossible," it would say. Later, the emerging caterpillar of many colors must seem to himself to be the pinnacle of remarkable life—how pretty with his many shades of green and yellow—and then he gets hungry and starts to eat. The miracle has only just begun.

We are told that the caterpillar can grow up to a hundred times his original size. As he molts out of the old skin and grows into a newer and larger one, he must be very satisfied with his improvement. Bigger, more colorful, more mobile—and then—something profound happens. By design of a higher power than the caterpillar, the two-inch long worm begins to weave or spin or otherwise create a strong, protective home where he or she will hibernate for a pre-designated length of time. By some remarkable—or even miraculous—happening, he finds himself encased in a strong, compact home where he will quietly reside for a pre-determined

spate of time, during which more miracles are taking place. Cell repro-
duction of the most complex sort is beginning to create eyes, antennae,
legs—and, of all things—wings! And beautiful colors specially fitted to
the unique species to which this creature belongs!

When the entire cycle is nearing completion and the final moment of
exquisite birth is fast approaching, the chrysalis starts to crack. Imagine,
if you will, the feelings of this living thing, unaware of its genesis or
even its purpose, as it emerges into a strange new world, crawling out of
a cocoon that it does not recall building. Finding itself hanging by one
silken thread to a twig high above the ground, struggling to free itself from
a small, unfamiliar oval box with appendages above and around him that
seem unwieldy but also seem to have a mobility which is coaxing him to
try them out—to fly?

To experience such a miracle as this newly-born butterfly must be
an astounding adventure, and more exciting than words can tell—if he
could think, or reason, or speak, or understand. But he can't. He can only
follow the urges that drive him. Also an incredible metaphor of our own
existence, if you stop to imagine—to allow things greater than our own
paltry understanding to take on new meaning as they unfold. New ways
of thinking can be intimidating, of course, depending upon our experience
or training—but fresh views are never wrong, if they are thoughtfully
accepted and examined before Yahweh Himself.

This is the template for building a construct where people of all walks
can take their inmost thought and new ideas and explore them with their
own personal "Jana"—that unique inner voice with whom one can explore
things not yet spoken—even to themselves. Did God, then, deliberately
give us the butterfly for a dual reason—to enjoy **and** to use as an example
of what He has for us all, as we allow?

Act 1 — Scene III

So it was when the first exploration of the language which existed between Yahweh and His people, and its deeper meanings, was initiated. So it was when, as a young child, I asked the first questions about music and its real meaning—and where it **really** had its genesis? Later, many years later or, maybe, just a few weeks later, one is surprised by new epiphanies which so often blossom into great bouquets of rare flowers. The heady fragrance of the flower permeates the air all around. The charge is not to waste the beauty of the epiphany, but to allow it to mature and learn from it. The epiphany of new insights or knowledge is a beautiful thing, to be handled carefully and thoughtfully, before casting it aside or recoiling from it.

Yahweh, the Almighty God, prepared a covenant to set before the Israelites, His chosen ones, to give them the rights and guidelines for a perfect life with Him. He also knew that their sins, committed over and over again, would repeatedly come between them and His perfect will for them. Nevertheless, as He had done in the past, He once more called them to put their faith and confidence in Him, and Him alone, and to join in that covenant with Him. Part of that perfect covenant was a language which He had, according to rabbinic knowledge and wisdom, specifically created for them—the Hebrew language and its accompanying numbers system that, at that time, were unexplored and uncharted.[26]

According to Genesis 11, the entire culture at that time enjoyed "one language and one speech" and was beginning to feel quite accomplished and powerful. The people said, *"Come let us build ourselves a city, and a tower whose top is in the heavens: let us make a name for ourselves."*[27]

---

26 Rabbi Mordechai Kraft: *The Hebrew Language is The DNA of Creation*: TorahAnytime: YouTube: August 31, 2013. https://www.youtube.com/watch?v=2j3OOxaQanw

27 Genesis 11:4

Yahweh was not pleased. The pride and arrogance of human kind has always been the wedge that blocks the freely offered mercy of Yahweh, and, in this instance, it was an obvious transgression that indicated that man had put his power to work to become as great as God Himself and to discover the genesis of that greatness. So, Yahweh said, *"Indeed the people are one and they all have one language, and this is what they begin to do; now nothing will be restrained from them, which they have imagined to do—Come, let Us go down and there confound their language, that they may not understand one another's speech."*[28]

And so it began. In order to short-circuit the growing power of mankind, facilitated by his expanding ability to converse and to understand the communications of everyone equally, many new languages were rather instantaneously born, much to the astonishment of the crowds of people gathered at the tower. Can you, in your sanctified imagination, just picture the horror when these power-drunk men realized that they could no longer understand each other, much less carry on any significant conversation with one another? If you would take this just a simple step further, remember that the first language was given under the caring hand of Yahweh, Almighty God, and He Himself was the Teacher of that language and the Source of learning which allowed His Israelite children to appropriate that language and speak it with fluency and understand it. Confusion abounded and, no wonder!

The ensuing ages only brought more branches to the tree and thereby further confounded the ability for men and women to enjoy communication with mutual understanding. A perplexing development, indeed. Hebrew, Arabic, Latin, Spanish, German, English, Gaelic, Farsi, Japanese,

28 Genesis 11:6-8

Chinese, Danish, Portuguese, Russian, Dutch, Indonesian; to say nothing of the various dialects in each of these languages! The fabric of linguistic communication was already being woven, and its evolution was, and remains, measured and veiled. Even as the pupa is in most ways unaware of its ultimate beauty and utility, mankind may quite well be in some ways oblivious of the beauty and power of the communication skills that come to him so naturally from birth. But that does not in any way diminish the dynamic utility of any language. It could, however, be an indicator that language is yet another common denominator underlying our irrefutable lineage that will eventually prove our glorious origins in the mind of Yahweh and spoken into being by His powerful Word—or song.

THE INTUITIVENESS OF LANGUAGE

The auspicious beginnings of language, if we are to accept the lore and wisdom of ancient rabbinical traditions, can inspire many beautiful flights of informative fancy. Again and again in this writing I shall allude to or call outright attention to the necessity for sound, and especially in the form of the human voice—or God's voice, as the case may be—as a cohesive force in creation. In order to follow the trail of my journey to find "truth in music" as God intended from those very beginnings of which the rabbinical scholars speak, one must grasp one cornerstone of thought: **sound is central to creation.** What sound? **The sound of the Word.**

Now the question becomes: What do we mean by the "Word"? It has to be understood that at the moment of final creation, God had already formulated and established His work **as** completed, as well as **how** He would complete it. True, it may have been incremental, sequential and deliberate—we will not know this side of the veil—but His design was complete at its inception. And He knew that the core raw materials, the core building blocks would amount to what physicists today are calling "frequency"

or "vibration" or "pitch." Scriptural references to the "Word" have gradually come to be understood as utterances from God; in fact, Jesus Christ has been given the name of the "Word of God Incarnate." The inferential connection is that words, or songs, or wave lengths, or vibrations are all linked by the marvelous mystery of "frequency"—or "sound."

Another mysterious consideration is the idea of energy—what is it and where does it come from? And how do we explain the fact that energy cannot be destroyed. Only by apprehending the idea of **sound** as a form—maybe **the definitive form**—of energy can we begin to catch a glimpse of what the scientists mean when they unequivocally state that "Vibration underpins all matter in the universe. No matter can exist without sound and vibration."[29]

By consulting with our sanctified imaginations and allowing our minds to venture beyond the acceptable bounds of how we think about language, sound and energy, we may be able to loosen new concepts from our preconceived notions and begin to imagine a beautiful new world that shimmers with vitality and life and is anything but static. "No matter can exist without sound and vibration." Does that possibly mean that words are living, moving things as well? Does that mean that the sounds of words have power that we have, heretofore, not recognized?

And later, we shall explore the idea that words and songs are parallel in their being, and that there is linguistic evidence to support that notion, as well as scientific experimentation.

All of this conjecture casts a new light on the whole subject of language and vocal communication. Words are not static and are certainly not limited in their scope. It does not take a deep investigation to trace

29 Cymascope: *Sound Made Visible*: Copyright 2008-2014 cymascope.com All Rights Reserved.

the similarities in different languages as we know them. Why is there such a strong similarity between them? Why do the same phrases carry the same meaning in so many differing dialects? The phrase "abracadabra," according to ancient Hebrew tradition, is derived from the ancient Hebrew language. The inference of the phrase is that "something is made from nothing by a word," which relates easily to the ancient story of the creation being spoken (sung) by God. And the phrase is common to all languages.[30]

The answer is only available to us if we are able to see the intuitiveness of our speech and language patterns. To understand this idea is to accomplish a new view of how and why we communicate in the first place and to see our vocalizations as more than just "words." The ways in which verbs express, the hidden nuances of meaning in our nouns and adjectives, the shape of our grammars are not accidental or incidental. They are deliberately planned shapes and nuances which are meant to carry much more meaning than we are apt to appropriate to them.

Language moves out of the mundane into the gloriously expressive and becomes a living entity—not fully understood.

WHAT CAN RABBINICAL WISDOM TEACH US?

It has taken more than forty years for me, with Jana's constant presence in my thought life, to come to terms with the complexity of insights that I am trying to express in this writing. With that in mind, perhaps it will be easier to understand how difficult it is to express those things as valid and even necessary to our grasp of the beauty and eternal scope of the creation of the universes. It is no small thing to begin to see

---

30   See JettieHarris.com.

that all things consist in the being and personhood of God, and that the cohesive element in that great miracle of creation is—**sound.** It is also mandatory that our minds be open to new ways of seeing the things that we already know. God's omniscience, for instance, is all-consuming: He knows **all** about **all.**

Early on in the development of this writing, I made it clear that the starting point had to be the acceptance of that Being, God Almighty—the reality of God as the Master-Musician, Master-Designer, Master-Planner, Master-Linguist, Master-Scientist; Master-of-All, in fact. Therefore, the thread of consistency is woven through every particle of every science, art or sub-creation that we can imagine. And so, it is very likely that the wisdom of the ages, as portrayed in the wisdom of the rabbinical writings and traditions should be a starting point for us—their knowledge was from the beginning of things and was kept by sacred purpose to be handed down from age to age and people to people. One does not have to be Jewish to benefit from the wisdom of the ages, but it is imperative that all people at least pay homage to the eternal sacredness of Jewish intuitiveness and knowledge.

My archetypical thinking was challenged by a lecture given several years ago by Uri Harel, a renowned Torah teacher from Haifa, Israel. From my youth to my old age, I had held the common belief that Genesis 1:1 said: *"In the beginning God created the heavens and the earth."* By simply explaining the Masoretic vowel and grammatical markings of the Biblical Hebrew text, Uri uncovered a whole new range of thinking for me by illustrating that the markings on the first letter in the original text directly changed the word "the" to the word "a"—"in a beginning."[31]

---

31 Hebrew: *Reading The Bible In Hebrew*: Genesis 1:1: https://www.youtube.com/watch?v=s6m5oy BY1uU Uri Harel: YouTube:August 31, 2013.

Act 1 — Scene III

Without embarking upon a deep theological discussion, suffice it to say that in the eternal scheme of things, God may have deliberately chosen to appoint several, or even, many, beginnings of differing import or function. Harel's position was that if God saw fit to arrange multiple "beginnings" we, as mere humankind, really have no right to question His omniscience.

Rabbinical tradition holds that the first creation was the formulation of the Hebrew alphabet of 22 letters, each of them having a specific meaning as well as a sound, and each of them bearing a numerical identity, as well. Out of this 22-letter structure, Yahweh spoke all of creation, including the rules of engagement in all of the disciplines to come, and the alphabet of the language was so perfect that it was able to provide a complete structure for all things. Then all the Creator God had to do was to continue speaking (or singing) words of His making in His language and the universes came into complete being.

Of, course, what this did was to open my mind to myriads of new possibilities of intuitive thinking and discourse with my companion and co-thinker, Jana, and enable me to become far less inflexible in my analysis of ideas.

In the course of my research, I became acquainted with the work of Rabbi Mordechai Kraft. Rabbi Kraft has served as the Assistant Rabbi at Congregation Havurat Yisrael in Forest Hills, NY for 15 years and is a popular teacher at Ateret Seminary, on campus at Queens College. He is also deeply involved in the work of EMET College Outreach, an organization that maintains an active affiliation for the discovery of and propagation of **truth** at eight New York City colleges: Queens College, St. John's University, LIU (Brooklyn campus), Baruch College, Hunter College, Queensborough Community College, York College and LaGuardia Community College.

The quest for **truth** has many faces—but the astounding thing is that **truth is truth,** and it is **irrevocably constant.** That is an amazing idea to finally come to terms with—all truth is immutable and related. My Hebrew professor at Liberty University told me years ago that to be able to understand "cross-disciplinary" relatedness was a gift—to be able to see that all things inter-connect.

Remember the butterfly chrysalis. Some species remain in the cocoon as long as two years without any knowledge of the processes going on therein. The final truth of the outcome is unknown to them. And so it is with us. Einstein was not able to solve the dichotomy between the macro- and micro-worlds, but the process was going on unbeknown anyway. And still is.

Brian Greene, math and physics professor from Columbia University, New York City, made this observation:

> "And so we see that the incompatibility between general relativity and quantum mechanics becomes apparent— That is, until the discovery of superstring theory—With the discovery of superstring theory, musical metaphors take on a startling reality. The winds of change, according to (this) superstring theory, gust through an Aeolian universe."[32]

This is only by way of indicating the verity of the statement, "Truth is truth." We may not be able to see the reasons why or the final out- comes just yet, but it remains a worthy consideration: all things intercon-

---

32 *The Elegant Universe*: Brian Greene: W.W. Norton & Company, Inc.: 1990> pgs. 130, 131.

nect. The issues of string theory and its related issues will be explored in another chapter.

The thread of thought, the "half-idea"—that all things are interconnected—has been key to my own quest, but I needed to know why. And then I made a most astounding discovery—the ancient wisdom of the rabbinical mind which knows the beginning of beginnings and can substantiate it with an in-depth study of an astounding entity—the ancient Hebrew language!

My field of expertise is music. Rabbi Kraft is a scholar of the Hebrew language and its inmost secretive revelations. He said this: (paraphrased)

> "Hebrew (language) defines reality. Here is a comparison—the chemical language defines reality from within. $H2O$ is water—two molecules of hydrogen and one molecule of oxygen. Look at the symbols and you can define the reality—change anything and you no longer have water. All other languages (of the world) are symbolic in nature; they have no intrinsic value (in their words). Hebrew is the **language of holiness**."[33]

I believe that herein lies a mystery which the Gentile mind, unless it is prompted by some outside inspiration or drive, has no way of understanding without making a deliberate study of the ancient Israelitish wisdom. The language itself is unique in a way that the word "unique" does not come near describing. The overview of the book, *Lashon HaKodesh: History, Holiness and Hebrew* by Reuven Chaim Klein, opens with this short paragraph:

---

33 *Rabbi Mordechai Kraft Secrets of Hebrew Letters*: http://www.esotericonline.net/video/rabbi-mordechai-kraft-secrets-of-hebrew-letters.

"Throughout Jewish literature, the Hebrew language is referred to as Lashon HaKodesh. Its history, origins, decline, and rebirth are simply fascinating. Furthermore, at its deepest level, Lashon HaKodesh is called such ('the Holy Language') because it is intrinsically sacred – and is thus unlike any other language known to man."

Can it really be that this mysterious, unexplored language holds secrets that until now have only been known by the scholars of the ancient Jewish lore—and, if so—why?

*"The secret things belong to the Lord our God, but those things which are revealed belong to us and to our children forever, that we may do all the **words** of this law."*
Deuteronomy 29:29

NOTE: VISIT http://JettieHarris.com FOR MORE INFORMATION

Scene IV

THE DUAL ROLE OF WORDS IN THE CREATION SAGA

> *Consider this…*
>
> **The incredibly intricate and well-orchestrated precision with which the universe functions could never be without a Designer.**

CONSULTING WITH JANA

Once upon a time, more than sixty years ago, there was a very dear friend—I shall call her Elle—who came to me in my senior year of high school. We shared ideas and thoughts that other fellow-students looked upon as so fantastic that they were almost thought to be crazy. Little did we care. We knew what we were thinking and doing and nothing else mattered that much. For me, this friend stood in for Jana for that beautiful period of time, short as it was. It was my senior year and her junior year. We climbed the West Virginia hills, taunted heavy thunderstorms, ran among the fields of flowers, roamed the streets of Fairmont in the rain and generally soaked in as much life as we could. For that period of time, I had no need for Jana; I had a real, live talking, singing friend with whom to share my deepest thoughts and questions. She visited me once after I left West Virginia, and then I never saw her again—for a very, very long time. Then, a year or two ago, I found her again. Now I have her AND Jana with whom to share thoughts that I cannot capture all by myself. How blessed can one human being be?

She has been listening to the ideas that Jana and I are crafting together as this chronicle unfolds, and, so often, she takes the nebulous, shapeless words and puts them in order. It is hard by email. But Elle sets things in an orderly fashion so that Jana and I can then expand upon them.

The task in Chapter III was to identify the depth and use of words as they are in truth, and not in supposition. Very difficult, because God is truth and simple words cast in the English vernacular fall so short. This is what Elle said:

"...readers who may regard the Bible as not only inerrant but also invariable are invited to consider two possibly radical ideas: (1) words (our human understanding) are social constructs with translations that can reveal profoundly different understandings and (2) words, the physical manifestations, are vibrations that may constitute the essence of creation."

And so, I bring this to the readers—consider that words are not merely meant to be social constructs. They MAY also be the very stuff of which the worlds are made—the holy vibrations of the voice of God as He shaped, first, the language by which He would create and then the energy of the vibrations which would be the very raw materials of creation itself. And who is to say that these words were not "sung"?

I must consult with my sanctified imagination, with Jana, to be able to follow this path, and I invite the reader to come along—

"God help me to clear my mind and with Jana's help, show me how to differentiate between the two..."

Beneath the Surface

"Half-ideas are transient-shaped,

Or else they must dissolve somehow each into each.

If only they would stand completely still

Until one found the words their size.

'I see, your measurements are thus and thus—

Act 1 — Scene IV

That's clear enough.'
You inventory your entire stock—
"Now this should fit'—and turn to find
It really doesn't fit at all.
You have a cubed suit
For a sphered thought.

You were sure that thought had corners."[34]

ne of the vicissitudes of life is the constant necessity to search and learn. Those who never learn that they must be incessantly striving to know, end their lives poorer. On the other hand, to be driven to learn is a difficulty, as well, because we sometimes don't even know what it is that we are seeking.

Very early on, I was captivated by the beauty of words and the power of using words—correctly or incorrectly. It soon became apparent that if I used the wrong words with the math teacher, the consequences could be long-standing. All through my school years I also noticed that for the most part, teachers and students alike seemed to take communicative skills very much for granted—except for a very tall, very thin Miss Bosch, the English teacher, who instilled a deep love for words in those of us who cared at all. She was often apt to wonder off the subject of the day in order to challenge us to think about where words had their origins and to question whether there was something deeper than the obvious hidden in the romance of Evangeline or the poetry of Kipling. It wasn't an obsession with her; just a rather casual inference that there might be more to language than what appeared on the surface. That made the omnipresent

34 *Dear People—Robert Shaw*: Joseph A. Musselman: Indiana University Press - 1979. pg. 72

memorization of declensions and verb forms more palatable—at least for me—because the suggestion was that there may be very good reason for them. Hard for a seventh grader to assimilate.

Parallel to this curiosity about language was the same curiosity about music. I had shown unusual talent for music very early in my life, but it was not possible for my parents to understand or to pursue developing that talent. My heritage as the eldest child of Appalachian parents also included a somewhat narrow-minded concept of fundamentalist Christianity. The good part of that was that, at least, I had a strong foundation of exposure to the Bible. By the time I was 10 or 12 years of age, I had memorized dozens of Scripture verses and one of the main themes was the representation of Jesus Christ as the Word of God.[35] Why the Word of God? Wasn't it enough to see Him as the Son of God?

"Nothing Comes From Nothing"—so goes the song from *The Sound of Music*. So—it must be that my childhood set the stage for the inquiries that were to come many years later. I could never let things lie. I had to find out why things were. How and why things were what they were. What the real meaning behind the obvious could be. And so, because I could not take advantage of higher education (that may not have offered the answers I needed anyway), I was channeled into my imaginative world where imaginary friends were prolific thinkers, and I could match wits with them about whatever things came to mind. Jana is still a valuable resource for me. Would that more numerous readers could find their own inner voice with whom to confer.

One of the things that interested me most was the idea presented in the Bible that Jesus was the "Word of God." This precept is easily grasped

---

35   John 1:1-3 New Testament: NKJV

theologically, I suppose, but in my inquisitive mind, I perceived that there could be some deeper implication. Seen in the light of my growing suspicion that the world of words is actually more than a simple means of communication, taking this statement that Jesus was the "Word of God" as a single edged thought left too many dangling threads of possibility. There had to be a reason for the Holy Bible to describe Him as such, and to leave that reason unexamined seemed confusing to me. To take any of the references to words and thought casually in Holy Writ seemed to invite lack of understanding of, not only **that** reference, but also of other references, which may be even remotely related or attached. And it was decades before I barely began to understand these implications: college, finally, at age forty and forty more years after that before my thoughts and "transient half-ideas" took shape to the point that I could express them.

LANGUAGE, WORDS, SOCIAL CONSTRUCTS

"Feathers in the wind." Folklore is replete with stories illustrating the power of the spoken word and its far-reaching consequences. Just try to even **think** about chasing after a pillow full of goose feathers which has been ripped open and the contents sent flying. Impossible! Daunting!

Mankind is wont to take for granted those things that present themselves as a natural part of life. Speech is one of those things that we can't recall being without. It comes to us almost as if by magic, and the child removed from his native America and transplanted into a new culture of, say, Germany, before the age of one internalizes the new language without any effort and later speaks it with the acuity of a native-born German. English-speaking families can expose their toddlers to any other language and fully expect them to assimilate that language with ease. Noam Chomsky believes that language skills are part of our hard-wired abilities that are inherent to our humanity. Take note:

51

"—certain linguistic structures which children use so accurately must be already imprinted on the child's mind. Chomsky believes that every child has a 'language acquisition device' or LAD which encodes the major principles of a language and its grammatical structures into the child's brain. Children have then only to learn new vocabulary and apply the syntactic structures from the LAD to form sentences."[36]

Because our language skills begin to develop so early in life, we may tend to take for granted that the only purpose for speech is to communicate and to get what we want or need for our own well-being. The child learning to talk is a hugely vain and self-centered human being. Satisfaction of his basic needs is the reason for his learning a new word. If he does not like peas, he very quickly masters the word "NO!" If he does not want his sister to play with his new truck, it becomes not only easy, but also necessary for him to learn to say "Mine!"

Refinements in expression come in time. Grammar is learned. Longer and longer sentences eventually become common and easily shaped. Some children become quite adept at framing intelligent explanations and even whole stories. Chances are, however, that even the most talkative child does not bother to analyze how and why he speaks the way that he does. Using and abusing language comes later.

Mountains of paper, gallons of ink, thousands of blogger's words and multiplied hours of experimentation and study have sought to unravel the wonders of spoken communication and the ways in which we learn

36 https://aggslanguage.wordpress.com/chomsky/

our verbal skills. To say nothing of the amount of ivory tower contemplation and discussion in which the scholars of such disciplines seek to give some sort of shape to the existence of language and its implications. Then, there are the vast hordes of people who never give a thought to the hidden, veiled secrets of language and its origins and evolution. Whether we pay close attention or whether we do not, the mysteries of language are captivating.

Did you ever stop to think that you are male or female because of the linguistic intents of some physician or midwife somewhere who was the first to see you?

Not that I believe for one moment that gender is a result of "social constructs"[37] devised by a linguistic necessity. But I do fear that the 20th and 21st century quibbles over gender and transgender have been deeply influenced by the presence of this thing that we call "social construct" which is a direct result of the function of language as a limited means of real communication. However, that is what we have. We are limited to the qualitative understanding that we have of our language skills, unless

---

37 Social depending construction is something you might not be aware of. You are somewhat living in segregation on what gender, race and class you are. Race, class and gender don't really mean anything. They only have a meaning because society gives them a meaning. Social construction is how society groups people and how it privileges certain groups over others. For example, you are a woman or a man because society tells you that you are, not because you choose to be. Simple as that. Just like it tells you what race you're classified as and what social class you belong in. It is all just a social process that makes us differentiate between what's "normal" and what's not "normal." According to the author of "Night To His Day: The Social Construction of Gender," Judith Lorber, the social construction of gender begins "with the assignment to a sex category on the basis of what the genitalia look like at birth" (55). When a baby is born, the first thing a doctor does is look at the baby's genitalia in order to determine whether it will be a boy or a girl; this is the beginning of the gender process of social construction. After they are classified as boy or girl, parents become part of this societal process as they start dressing them with colors that identify their gender. The "normal" thing to do in this case would be for baby girls to be dressed in pink and baby boys to be dressed in blue. It is just not normal to dress your baby boy in pink or your baby girl in blue, right?

http://oakes.ucsc.edu/academics/Core%20Course/oakes-core-awards-2012/laura-flores.html

we give special attention to them. Our curiosity about language is essentially a subjective matter. Not interested? No matter. Just learn enough to make your needs known. But I believe that there is much more than this to be appropriated that could improve our lot in life and our ability to interact and raise our standards higher. Pitfalls of interpretation abound. But careful attention to the great gift of speech and its ramifications has the potential to frame a better social environment.

There is little argument over the fact that words are incredibly powerful, for good or evil. But the extent of that power to form and teach could be debated. There is an increasing number of scholars, however, who are giving strong attention to discovering the veiled ramifications of language. In their fascinating book, *Words Can Change Your Brain*, Andrew Newberg and Mark Robert Waldman, make this observation:

> "—a single word has the power to influence the expression
> of genes that regulate physical and emotional stress."[38]

Theresa J. Borchard writes an interesting commentary on this book by Newberg and Waldman:

> "Positive words, such as 'peace' and 'love,' can alter the expression of genes, strengthening areas in our frontal lobes and promoting the brain's cognitive functioning. They propel the motivational centers of the brain into action, according to the authors, and build resiliency.

---

38 *Words Can Change Your Brain*: Andrew Newberg and Mark Robert Waldman: Hudson Street Press: Penguin Group, Inc.; 375 Hudson Street, NY NY 10014: USA: 2012.

"Conversely, hostile language can disrupt specific genes that play a key part in the production of neuro-chemicals that protect us from stress. Humans are hardwired to worry—part of our primal brains protecting us from threats to our survival—so our thoughts naturally go here first."[39]

Viewing the marvel of human speech and vocal communicative skills is fascinating enough in its own right, but when you take the time to consider, as Jana and I have considered, the possible connections between the miracle of human language and its spiritual counterpart, the wonder of it all compounds exponentially.

Of great interest to me in the continued development of this book is the incredible insight of so many scholars in so many differing disciplines, as well as the depth of knowledge from which they approach their various works. And yet, for all of it, they fail to grasp the connectedness and the vast extent of what they have discovered. For instance, re-read Borchard's commentary on *Words Can Change Your Brain*.

"Positive words, such as 'peace' and 'love,' can alter the expression of genes, strengthening areas in our frontal lobes and promoting the brain's cognitive functioning. They propel the motivational centers of the brain into action, according to the authors, and build resiliency.

"Conversely, hostile language can disrupt specific genes that play a key part in the production of neuro-chemi-

---

39 http://psychcentral.com/blog/archives/2013/11/30/words-can-change-your-brain/

cals that protect us from stress. Humans are hardwired to worry—part of our primal brains protecting us from threats to our survival—so our thoughts naturally go here first."

It is not difficult for me to infer that behind all of our plethora of technical, scientific, artistic and historical information is a commonality that can only point in one direction—to a unified and perfectly designed source. The only explanation for that source is that it must be of an infinite nature and, therefore, we can infer, as well, that it must be of some level of ability far, far above that of mortal, finite man. It seems only reasonable to me that this "transient idea" should be investigated before mankind makes the conclusions that he is wont to make—i.e., that all things come from the intelligence and accomplishment of mankind.

This is the complexion of the thinking that one can only have within oneself in order to effect incredible changes, Those changes could be life-altering.

I have personally seen Theresa Borchard's ideas about life-changing language work in an astounding way. Hostility and confrontation were the ground-work of my family for generations. I deliberately began to alter my communications with one family member several months ago to a level of kindness and acceptance that was, at times, difficult to maintain. In a short time, I began to hear humor and smiles creep into the conversations. I can safely say that, at this time, our conversations are almost completely devoid of that generational hostility and unkindness. For us to confer with our inner selves and explore new thought is becoming more and more vital. Consider the inherent weight and power of words, and what the possible connection with infinite God might be.

Act 1 — Scene IV

Behind the overt motivation for this book is the idea of "beginnings" for all things—creation, itself, and the miracle of how all things are related. To see all things as related becomes a simpler task, if one grasps one fundamental thought: All things would have to be related if, indeed, they were designed, put into motion and built into reality by the creative and inventive power of one Person: Yahweh—God Almighty. As the Omniscient, Omnipresent Designer-Creator, He would, by His own nature, have to have a completed and perfect plan in place before the actual act of creation took place. This thought in and of itself is quite beyond our comprehension, all the more so if we do not necessarily base our entire world view on the holy and sacrosanct Judeo-Christian Holy Writings: Torah, Talmud and the Holy Bible.

Earlier in this writing, I called attention to the idea of the physical world being an analogy or metaphor for the spiritual world. This is a "transient idea"[40] that I have carried in my sanctified imagination for many years along with the idea that all things are related in some way. The incredibly intricate and, yet, incredibly well-orchestrated precision with which the universe functions always held a secondary wonder for me: nothing this intricate and obviously perfectly organized could be without a Designer Who had, has and will have in the future, omniscient power to do and be. Everything in nature seems to cry out "faultless design." The problem is that there is and was no empirical proof of anything that I was thinking. And, of course, the temporal world demands proof, doesn't it? How could this "transient idea" find its well-fitted suit?

NOTE: VISIT http://JettieHarris.com FOR MORE INFORMATION

---

40 *Dear People—Robert Shaw*: Joseph A. Musselman: Indiana University Press - 1979. pg. 72

Scene V

THE END OF THE BEGINNING

> *Consider this…*
>
> ## When Scripture says that God created out of nothing, that is precisely what it means. There was nothing, and then—GOD SANG!

I turn to Jana, with God's help to weave together the threads of knowledge that I have gathered over the years and make them into the rich and colorful fabric that I see in my sanctified imagination. I see an unending universe, always expanding into more and more beauty as it runs out into infinite space. I hear a symphony that fills every star, every planet with such ethereal music that the human part of me cannot decipher its beauty. I see a blue planet far away, white clouds swirl like a halo all around her body—it must be my Earth—the place where I grew to old age—and now so far away and yet so near. Everything that I am seeing and hearing now is magnified in glory and I must wonder where I am and where I am going. I know—it is a brief preview of the glory that is to come when our God makes all things known to us and we will know as we are known—is this "emet"—the ancient Hebrew word for "truth"? Is this a glimpse of the "truth" expressed in הקודש לשון, LaShon HaKodesh, the Holy Language? Is this the song we will sing?

Give unto the LORD the glory due to His name;
Worship the LORD in the beauty of holiness*. Psalm 29:2*

59

God's Opera

*H*ow incredible it was to my thinking when I discovered the work of two Hebrew scholars: Uri Harel, founder of the Center For Biblical Hebrew in Phoenix, Arizona, and Rabbi Mordechai Kraft, an exceptionally gifted lecturer on the secrets of the ancient Hebrew language.

Uri Harel was a native of Haifa, Israel, educated there and in Buffalo, New York, before moving to Phoenix where he spent a large portion of his life teaching and lecturing. The main emphasis in his work involved his deep interest in the musical aspects of the Hebrew language and the effects of music on the human organism. But he was also greatly gifted in teaching the rich nuances in the ancient Hebrew language as well as the basics, which are so necessary to a fine understanding of that language.

His untimely death from cancer in 2012 left his work to be carried on by some of his students and colleagues. One of them, Dr. Anne Borik, of Phoenix, is still active in pursuing his work and is also doing research on the effects of sound and movement as therapy for many different physical challenges. Her research and some of her conclusions will be addressed in the ACT of this book devoted to exploring music.

Rabbi Kraft has been Assistant Rabbi at Congregation Havurat Yisrael in Forest Hills for 15 years and is, as mentioned in a previous scene, one of the lead staff members of *Emet* Outreach, an active college outreach program with energetic presences on eight New York area campuses. His lectures for the website TorahAnytime, Inc.[41] are exciting, informative and spiritually enlightening, and, I believe, timely in the sense that there are truths buried in the ancient Hebrew traditions and language that he and other scholars of the Torah and Kabbala understand and are willing to share with the world. And, I believe that it is exciting virgin information for many of us.

41 © Copyright TorahAnytime.com 2006 - 2014. All Rights Reserved

Act 1 — Scene V

The Hebrew word, אמת (*emet*), means truth, and it is interesting to me that the study of the ancient Hebrew tradition, as taught by these two brilliant men, is so closely related to the things that I have been thinking for so long, and yet have, up to now, been unable to decipher or research properly. Why? Because much of what I have been thinking has been intuitive—based on study, prayer and knowledge of other disciplines—but basically intuitive. I categorize my work as being intuitive because I have been deeply engaged in melding together accepted knowledge from many differing and often, adverse, theories and academically proven areas of study. It is, presently, a widely accepted idea in academia that inter-disciplinary thinking and teaching is profitable. Is this a prelude to accepting the idea that all things may, in fact, be related—as if all things were, indeed, designed by one Creator and Designer? I see no conflict, actually, between theology and science or between art and science or between math and music. In my sanctified imagination, each one of these specific areas of knowledge and expertise graphically complement each other.

Ten years ago, little did I suspect that my quest for "truth" would find its fruition in the study of Hebrew and the ancient Hebraic tradition. It is beginning to appear more and more obvious that, indeed, the loose ends of my imaginative and intellectual roamings are almost exclusively going to be bound up in the wisdom of the ancient scholars and their intimate involvement with Yahweh and His holy language, הקודש לשון. (L' shon HaKodesh)

The intriguing words, *"The secret things belong to the Lord, our God, but those things which are revealed belong to us and to our children forever, that we may do all the words of this law,"*[42] cause me to ask, "Is Yahweh now beginning to open doors of understanding that will change the direction of destiny in our world, as we humans now see it?"

---

42 Deuteronomy 29:29: *The Holy Bible, New King James Version*: Copyright 1982 by Thomas Nelson, Inc.

In his incisive book, *The Day the Universe Changed*, James Burke made the statement: "...We are what we know—and as the body of knowledge changes—we also change."[43] The final and omniscient Keeper of all knowledge of all things is the Creator God—Yahweh—and only to Him does the "body of **all knowledge**" belong. Therefore, although what Burke says concerning the given body of knowledge at any particular time in the history of man is true to a certain extent, all knowledge and all dissemination of that knowledge belongs alone to Creator God—and how and when He prefers to allow us to "know what we know." It is rather a source of amusement for me to think of my God, Who is the Author and Finisher of all things in my mind, to be sitting on high doling out little bits of "knowing" as He sees that we can assimilate it and—maybe—even use it to our own good and His glory.

The whole idea of creation has been a mystery and a matter of fierce discord and disagreement for hundreds of years. Could it be possible that God, in His infinite wisdom is just now opening the windows of new consideration so that we enlightened people wandering around on this beautiful globe that we call home can begin to see inside the citadels of learning and get a fresh glimpse of the realities of creation? Consider.

"A" BEGINNING
הַשָּׁמַיִם את אלהִם בָּרָא בְּרֵשׁת

New knowledge is an exciting thing, and especially when it helps to clarify, to explain, to enlighten one about something that has been problematical or has created a barrier to learning of some sort. Of course, in order to be an unrestricted receptacle for new knowledge, one must be committed to engaging new thought without prejudice. And that is not a

---

43   James Burke: *The Day the Universe Changed*: Little, Brown and Company: September 1986.

simple task. One must have a sanctified imagination, free to think, in the vernacular, "outside the box," and to permit unconventional, maybe even capricious ideas to live long enough to take shape and find fortification in **fact**, if, indeed, these new ideas are worthy of finding that well-fitting suit of clothes that Robert Shaw was so fond of talking about. It could be a very good thing to find one's "Jana" and really inspect new ideas for their own unfettered worth with the help of an alter-ego or inner confidante.

As a music school undergrad, I first began thinking about the **origins of music**, only to find myself then beginning to question **origins in general**, and I began to intuit a strong connection between music, theology and creation. The trouble was that I could find no research materials to speak of that would integrate information related to all three of these disciplines. And, there was a fourth consideration, science. The basic tenet of science is empirical proof. Or so, the accepted premise is. To find a suitable garment to fit all of these varying disciplines was not even plausible.

The writings of one scholar from the U.K., Jeremy Begbie, proved to be invaluable, but, here again, his path was somewhat different from mine. His book, *Theology, Music and Time*, became my handbook and veritable bible, and he selflessly and eagerly conferred with me and has continued to encourage me to "think." And then, there are the writings of J.R.R. Tolkien and C.S. Lewis which offered ideas about the singing creator gods and the evil counterparts, but which could not be considered as research because they were, in fact, writings of a fantastic bent—allegory.[44]

It was, for me, the beginning of an arduous quest of learning about one subject, then another and trying to make my own decisions about the connections and integrations. I had studied several romantic languages

---

44   C.S. Lewis: *The Magician's Nephew* and J.R.R. Tolkien: *The Silmarillion.*

as a voice major in college, but the study of Hebrew was not even on my radar screen. Biblical commentaries, concordances, and theological treatises gave me very little help. Even three years of classic Hebrew study at Liberty University did not openly give me the insights that I wanted— maybe because at that time, I was not sure what I was really looking for. Of course, the knowledge I did gain made it possible for me to understand what was to come when I finally stumbled onto the teachings of Uri Harel and Rabbi Mordechai Kraft.

Information was laboriously gathered, categorized and, then, even more laboriously thought through, so that it could become an integral part of my growing theory about origins and beginnings—creation. The development of the idea that creation was, and continues to be, a matter of musical importance has grown slowly over more than forty-five years. In fact, it could be said that this entire work has been a life-time event, because my first contacts with music began when I was a mere five or six years of age. And it is out of those early beginnings that the intensity of my quest has grown.

It is still burnt deeply into my memory—that fateful evening several years ago, when I was so mysteriously led to PBS to watch Brian Greene, of whom I had not heard before that evening. Greene is the Co-Founder and Director of Columbia University's Institute for Strings, Cosmology and AstroParticle Physics, a research center seeking implications for theories of cosmology. His work, embodied in his seminal book, *The Elegant Universe*, was my first glimpse of something that I had "intuited" for all of my thinking life-span: all that exists here on earth and in the far-reaching spans of space is somehow comprised of, held together by and given reason by the reality of **vibrating energy, which is the essence of music**.[45]

45   *The Elegant Universe*: Brian Greene: W.W. Norton & Company, Inc.: 1999.

Act 1 — Scene V

Several years later, by the same workings of mysterious curiosity, which I recognize as the definite leading of God granting me new insights, I was brought into contact with a new and emerging science called "cymatics." My son, whose mind is even more inquisitive, maybe, than my own, called several months ago to tell me about a new website concerned with this emerging science.[46] The basic premise of these men and women studying "modal phenomena" is that "vibration underpins all matter in the universe. No matter can exist without sound and vibration."[47]

Encased in the knowledge of string theory and cymatics is a concept that is all important to grasping the idea of a dynamic musical aspect in the creation chronicle. First of all, remember that music is the only art form, which demands a time frame as an essential. Also remember that time is an essential in the world of mankind—not in the world of God. However, our Creator uses time and all of its concomitant concepts to the advantage of His primary creatures—mankind. He gives us myriads of examples by which to understand and profit from the concept of time. Therefore, I feel free to assume that music may be considered as a good analogy to the creation saga. It is ongoing. It is continually creative. It is forever changing. It has beauty as one of its most vital component parts.

Both string theory and cymatics lay claim to vibrating energy as the cohesive and adhesive element in the creation and maintenance of the universe. When the "vibrating energy" of a symphony ceases, the symphony is **over**, and the art work is finished. That is, it is finished as far as we humans beings can perceive, but well-known conductor Robert Shaw had the distinct spiritual/philosophical belief that sound, once energized, had no end—it kept going into infinity. He may have been exactly right.

---

46   Copyright 2008-2015. cymascope.com. All Rights Reserved.

47   Welcome to the home of the Cymascope: *Sound Made Visible*: 2008-2015.

On the other hand, the creation symphony **is** ongoing, and can be proven scientifically to be a perpetual event. The galaxies are expanding and rushing away from the earthly tether at an incomprehensible rate, and science is prepared to document that fact.

Consider the tenet presented by Uri Harel in a series of lectures,[48] "Hidden in the Hebrew," given at the Center For Biblical Hebrew, Phoenix, Arizona, before his death in 2012. Based on his broad knowledge of the rabbinical traditions of ancient Israel and the Biblical Hebrew language, he believed that the creation chronicle was intended to be and, indeed, continues to be dynamically manifest in 2016. This is an exciting affirmation of one my basic beliefs—that all things in creation, being inspired, generated and maintained by the Omniscient Almighty God—are related at the most infinite and inexplicable level.

Harel makes a simple and yet profound reference in illustrating that creation must be a perpetually active concept: You have just bought a beautiful, new car and you have driven it out to the desert and parked it. Years later, you go out to the desert to visit your beautiful car. It is no surprise that you find it in a condition of decay and ruin. Anything created and left unattended will eventually be destroyed, simply by the forces of nature themselves. He suggests that this is also a possibility in the vast cosmic world; it must be continually renewed and the laws of creativity are in effect to provide for that need of continual renewal.[49]

This idea is supported graphically by Scripture. It is not difficult for people of evangelical background to understand the concept of Jesus Christ as Co-Creator, but it may afford a troubling thought for others,

---

48 Genesis 1:1—Hidden In The Hebrew: Uri Harel: Lecture #47: YouTube.

49 Lecture 47—Hidden In The Hebrew: Uri Harrel: YouTube.

being of a differing predilection. The Book of Colossians is quite clear in its presentation of Jesus as just such a Co-Creator:

*"He is the image of the invisible God, the firstborn over all **creation**. For by Him all things were created that are in heaven and that are on earth, visible and invisible—All things were created through Him and for Him. And He is before all things, and **in Him all things consist**."*[50]

Robert Shaw used the metaphor of the "transient-shaped idea" often. He called it a "half-idea" and the inference was that so many, many thoughts and ideas had so many, many component parts and were, by their very nature, impossible to pin down and could, also by inference, cause great consternation and, maybe, even doubt. But he **never** demurred from a new thought or idea until he had thoroughly turned it over and over and had diligently tried to (find) the words to fit the "size" of the thought or idea. It was this process of deep introspection and analysis that afforded him the ability to continue growing and learning until his inevitable death in 1999.

The essence of this **"transient-shaped idea"** will connect God and Jesus Christ to the process of speaking or singing and thus will establish the possibility of creation as a musical event. And it may bring the reader head-on with another difficult proposition: The idea is that one must reconcile the thought that the first sounds ever heard in the eternal fabric of what was before, what was during and what came after the "creation of the universe" were sounds that emanated from the heart and voice of Almighty God. And whether those sounds were sudden and all-encompassing or were, on the other hand, like a symphony composed or a sacred oratorio sung, they were produced by the intent and energy of YHWH

---

50   Colossians 1:15-17: New King James Version: 1990.

Himself, and all the emanations of whatever end, were by His design and intent entirely. And they included all of the laws and rules and regulations needed to proliferate all of life and activity until the end of time as we know it. They included all of the harmonies, timbres, rhythms, melodies and pitches necessary to provide eternal creative possibilities to keep the cosmos in its course until that critical time of apocalyptic transformation into a "new heaven and new earth."

The Torah, as the original Sacred Writing, which has become our Old Testament, clearly states that YHWH was the Author of the creation story and that it was the sounds of His voice that *"created out of nothing."* The word *barah* carries the veiled meaning of just that—to speak (or sing) and by the power and energy of that vibration, the creation of all that is becomes reality. In the second verse of the Book of Beginnings, the Hebrew word which has been translated as "hovering," (*racaph*) brings a baffling series of possibilities to our attention: It has been translated as 1) to hover, 2) to relax, 3) to shake, 4) to tremble. I have even found one translation which refers to the "hovering" or "trembling" as "vibrating." That idea is curiously consistent with the thinking of the string theorists and the cymatists that it is, literally, vibrating energy or vibrating "strings" of energy that maintain the cosmos in its orbits.

Uri Harel points out that it does the reader well to refer to several English translations. He points out an interesting detail in *Young's Literal Translation* that states in the first verse of Genesis, *"In the beginning of God's preparing,"* which may alleviate some of the conflicting thinking about the first moments of creation.

All of this multitude of words says one thing: We are faced with a subject too far reaching to solve in one volume. Therefore, again I say, this book is meant to inspire other thinkers and scholars to pursue what catches their own interest the most. This is a volume of "transient-shaped ideas," and although all of them are based on prayerful research and many years of consideration, they are not meant to be taken as fact, or even, theory—

The prominent idea that I wish to leave with my readers in this ACT which deals with the sounds, frequencies, vibrations and pitches which I call the "Song of God," is simply this: As many faceted as it is, the idea can be reduced to the possibility that creation came about as a result of the singing of Almighty God, as He formulated the language, sounds, figures and energies that could provide for the ultimate language of musical eventualities. When Scripture says that He created out of nothing, that is precisely what it means: there was nothing, and then—**GOD SANG.**

The leap of faith pressed onto us in this process of sanctified imagining is to wipe away other theories about this heavily convoluted subject, at least temporarily until we have had time to find new "words the size of the new thoughts."

SOUND & THE SCIENCE
BEHIND THAT SOUND

הָאָרֶץ וְאֵת הַשָּׁמַיִם אֵת אֱלֹהִים בָּרָא בְּרֵאשִׁית

Scene I

GOD'S SONG—THE ORIGINAL SOUND

> Consider this…
>
> *If you can find enough freedom of thought to comprehend total silence and void, then imagine an explosion of glorious, all-encompassing sound that would defy all other possible sounds.*

To ask the reader to take seriously the ideas presented in this book and yet to allow the sanctified imagination to fly may seem like a profound contradiction in intellectual acuities. Not so. So often, our intellectual acuity can become more constricting than liberating. The study of Biblical Hebrew opened doors and windows of knowing that I had no idea even existed. And how exciting! How exciting to find nuggets of knowledge tucked away in a beautifully alive language that literally have been life and thought altering.

*As a new student of Biblical Hebrew at Liberty University in Lynchburg, Virginia, in the early 2000's, I was struck by a thought that I had no real basis for, but I could not wipe it out of my mind: **Ancient Hebrew is a living language**. It was one of those intuitive "transient-shaped thoughts" that have been my hallmark for decades, and now seem to be standing me in good stead as I formulate this new conjecture, if you will, about the creation saga. Little did I know that as I progressed in my*

73

*own research I would be learning probably far more than I would have
the ability to express and at the same time generate the excitement and
profound joy that each new discovery gives—*[51]

THE GENESIS OF SOUND AND
THE WORK OF CREATION

cience has been searching and speculating over the issue of
creation for eons of time. The origin of matter, the nature
of matter, the building blocks of matter have been the subjects of supposition and theorizing ever since scientific thought came
into being. The mechanics of the universe, the mechanics of time, the
mechanics of space have been the focal points of queries of all kinds ever
since the study of physics surfaced. The similarity of one discipline in
relationship to another has had a definite impact on how these ideas interact. Because most scholars, due to the intensity of focus demanded by
doing any research, seem to become quite exclusive when dealing with
their separate disciplines—to the exclusion of the beauty of allowing one
idea to directly relate to and influence the formulation of another from
another discipline or point of view. What I am asking the reader to do is,
in the sanctity of their own private and personal imaginations, to allow
themselves to conjure up how, for instance, the science of sound could
possibly interact with, say, a theological idea about the building blocks
of creation and the possibility of those building blocks being musical in
nature. It is, indeed, a large leap of creative thought. And, in all probability, it will amount to being able to keep an open mind from the beginning
of this book to the very end in order to see the warp and woof of beauty

---

51 These are my own thoughts, which provide some sense of direction to the reader in terms of chronology,
as it could be difficult to maintain a stream of ideas when covering so many ideas at once otherwise.

that results from allowing the mind to embrace the notion of "related-ness"—true "relatedness" or "connectedness"—of the depth or caliber of Einstein's "one set of simple and beautiful laws" that would be able to explain the entire universe and its mechanisms.

There are several prevailing ideas that thread their way through all of the conjecture over the universe, matter, space, time and the mathematical accuracy exhibited and necessitated by the co-existence of all of these entities.

It may be helpful at this point to review what has been proposed as a starting place for this new view of the creative saga and its vital meaning to mankind and all of his concomitant involvements, activities and accomplishments. The baseline question may be: what actually is the essence and purpose of life and all of its corollaries and implications? Or, in the words of the *Shorter Westminster Catechism*—Q 9: "What is the work of creation?"[52]

"A. The work of creation is, God's making all things of **nothing**, by the **word** of His power, in the space of six days, and all very good."[53]

Indeed, the work of the Westminster Assembly is astute and spiritually sound, at least in the view of reformed theologians, but it is not assumed to be Holy Spirit inspired, and therefore remains in the realm of man's work and not necessarily God's. But it does give us a firm springboard from which to undertake the sanctified imagining of some sort of connection

---

52 The Westminster Shorter Catechism is a catechism written in 1646 and 1647 by the Westminster Assembly, a synod of English and Scottish theologians and laymen intended to bring the Church of England into greater conformity with the Church of Scotland. The assembly also produced the Westminster Confession of Faith and the Westminster Larger Catechism. A version without Scripture citations was completed on 25 November 1647 and presented to the Long Parliament, and Scripture citations were added on 14 April 1648.

53 Genesis 1:1. *In the beginning God created the heaven and the earth.* Psalm 33:6, 9. *By the word of the LORD were the heavens made; and all the host of them by the breath of his mouth.... For He spake, and it was done; He commanded, and it stood fast.* Hebrews 11:3. *Through faith we understand that the worlds were framed by the Word of God, so that things which are seen were not made of things which do appear.*

between the sound/science undergirding the creation saga and the equally firm foundation of the creation of the world emanating from the **"sound of the voice of God."** To make the leap from one to the other is no simple task, unless one is willing to exercise one's sanctified imagination to open the mind to the impressive notion of relatedness.

CONNECTING THE DOTS

This is probably a good point at which to review the ideas that gave birth to the theory that creation was and is a musical mega-event and that the building blocks of matter are musical/vibrational in nature. It is extremely difficult to follow the journey from inspiration to fruition when you are dealing with such a vast commentary as all of creation. In fact, I have had over forty years of growth and intellectual expansion in which to couch my ideas and my understanding of them. But to translate that journey of insight and development in such a way as to be able to make them clear and accessible to the reader is another issue—quite.

However, to make that attempt is imperative. The presence of music in the universe has been known and written about for centuries. The early Greeks believed in the "ethos of music," and their greatest philosophers conjectured that music of certain modes had a "deleterious effect upon man." Pythagoras wrote of the "music of the spheres," and, in fact, based his entire theory of numbers on musical understanding. Galileo was renowned as "one of the greatest scientists of all time," and it is probably not coincidental that his father was one of the most influential musicians of the late Renaissance. The constant presence of music in the young life of Galileo was undoubtedly a formative factor.

Indeed, it seems quite elemental to me that early intellectual journeys were always in some way shadowed by or enmeshed in musical considerations of one sort or another.

Act 2 — Scene 1

Another dot which must be connected to the fabric is the existence of the Biblical accounts of singing, dancing and various other musical endeavors. There was no written language for a time, and that necessitated the use of song to pass along cultural mores, religious teachings and news. There are many various, if subtle, references in the Bible to the ways in which music became warp and woof of the ancient Hebrew cultural landscape.

The development of the vocational and professional life of humankind is particularly interesting to me:

> *"Then Lamech took for himself two wives: the name of one was Adah, and the name of the second was Zillah. And Adah bore Jabal. He was the father of those who dwell in tents and have livestock.* **His brother's name was Jubal. He was the father of all those who play the harp and flute.** *And as for Zillah, she also bare Tubal-Caine an instructor of every craftsman in bronze and iron."*[54]

Think back to those undocumented days when everything under the sun was in the early process of coming into full fruition, and imagine how it must have been. No one knew how to play the flute or the harp. In fact, there were no flutes or harps! Therefore, it is not a huge leap of fantasy to realize that this Jubal-person must have also been the inventor of these wonderful things that were endowed with such beautiful capabilities and could make such exquisite sounds. And Who was the teacher of Jubal? Who gave him the understanding of melody and harmony and Who instructed him in how to build an instrument which could embrace

---

54   Genesis 4:19-22: New King James Version: 1983: Thomas Nelson Inc.

and express all of the beauty at Jubal's disposal? Imagine—the very first musical instrument constructed by the very first musician with which to sing the very first hymns of praise! It piques thoughts of the vastness of creative energy to consider it!

Eons went by. Curious and thinking people could not avoid exploring the world into which they had been propelled. As the numbers of people increased, so did the vastness of their explorations. There is neither time nor space in this volume to follow the ages of cultural, religious, academic, philosophical, scientific and musical development and expansion. But it is both interesting and, I believe, necessary, to allow time for the mind to go back to the very beginnings and meditate on the "way it was."

In essence, there was "nothing." The very initial act of creation, which was exclusively God's to be kept forever as an eternal happening, began in a "void." The Hebrew language contains strong inferences that Almighty God spoke (or sang) and that the very sound of His voice put into being the universes and all that is in them. This idea has already been approached in the first Act of *God's Opera* and will be further developed in the Act dealing with theology. Suffice it to say, though, that there is strong support for the "transient idea" that sound, and in particular, musical sound is an elemental dynamic in the quest for explaining and understanding our universe and all that is in it. And—the ancient Hebrew idea that God's voice is the source of the elements of which that creation has been made out of "nothing" is becoming more and more a viable idea. Hence, *God's Song* and *God's Opera*.

There is another thread of knowledge demanding to be considered as we continue to "connect the dots." The **science** of sound. If God's voice were the source of all sources, what does that mean to us as we seek to unravel the enigma of creation and its meaning and its work? How does that fragment of information play in terms of the empirical evidence that is necessary to the scientific mind?

In Wittenberg, Germany, in 1756, Ernst Chladni was born. He was to become known as the "father of acoustics." His was a rich and varied career, since he gained renown both as a physicist and as a fine musician. He did extensive research with the vibration of rigid plates upon which he scattered sand or salt and then excited the plates with his cello or violin bow. What he discovered was that differing and changing vibrations of the plates caused the particles to spontaneously dance into varying shapes, depending upon the frequencies of vibration. He also discovered resonant frequencies which caused areas of the plates to stop vibrating and others to vibrate at differing rates, setting up resonant patterns and non-vibrating of "silent" nodes where the particles ceased to move.[55]

From here, we can easily move into the 21st century to the remarkable research of a group of scientists from the UK, US and Denmark. They have expanded the research of Chladni to a vast degree and have broadened their outlook to include the healing properties of vibration and sound. That particular facet of their work does not affect the theoretical line of this work, but it is a vastly important emerging science that I believe will have powerful ramifications in many fields, not the least of which is the study of creation. This emerging science is "cymatics" and

---

55 One of Chladni's best-known achievements was inventing a technique to show the various modes of vibration of a rigid surface. When resonating, a plate or membrane is divided into regions that vibrate in opposite directions, bounded by lines where no vibration occurs (nodal lines). Chladni repeated the pioneering experiments of Robert Hooke who, on July 8, 1680, had observed the nodal patterns associated with the vibrations of glass plates. Hooke ran a violin bow along the edge of a plate covered with flour and saw the nodal patterns emerge. Chladni's technique, first published in 1787 in his book Entdeckungen über die Theorie des Klanges ("Discoveries in the Theory of Sound"), consisted of drawing a bow over a piece of metal whose surface was lightly covered with sand. The plate was bowed until it reached resonance, when the vibration causes the sand to move and concentrate along the nodal lines where the surface is still, outlining the nodal lines. The patterns formed by these lines are what are now called Chladni figures. Similar nodal patterns can also be found by assembling microscale materials on Faraday waves. From Wikipedia, the free encyclopedia.

is led by John Stuart Reid, supported by Jytte Brender McNair and Louis H Kauffman.[56] This team began its work in earnest in the 1990's and has added to their number since and extended their work significantly. The following statement set forth their stated mission quite well:

> "—we have remained aware that the governing dynamics of cymatic phenomena must be mathematically described for it to be fully embraced by mainstream science. In 2012 we began working with the team of mathematicians and scientists who, after discovering what may be **the energetic pattern that resides at the heart of creation**, have spent almost two decades investigating and researching its application. The anecdotal perspective of this discovery was first presented in the book *The Pattern* in 1997; the Pattern is now known as 'Mereon.' In our first dialogue with the Mereon team we were asked if the CymaScope could render visible some of the key frequencies related to this pattern. No one could have predicted just how successful the results of this collaboration would turn out."[57]

The work of this group of scientists and scholars is probably the most easily accessible in terms of understandability. They have, in my estima-

---

56 John Stuart Reid (b 1948) is an acoustics engineer who carried out cymatics research in the King's Chamber of the Great Pyramid of Egypt in 1997. Reid published his research results in Egyptian Sonics 17 containing photographs of the cymatic patterns that formed on a PVC membrane he stretched over the sarcophagus. The experiment was designed to study the resonant behaviour of the granite from which the sarcophagus is fashioned.

Jytte Brender McNair is presently associate research professor at the Institute of Health Science and Technology, Aalborg University, Aalborg, Denmark.

Louis Hirsch Kauffman is an American mathematician, topologist, and professor of Mathematics in the Department of Mathematics, Statistics, and Computer science at the University of Illinois at Chicago.

57 Cymatics.com. John Stuart Reid.

tion, taken the theoretical ideas of the string theorists and put them into credible visual form. I realize that from the academician's point of view, that may not be empirically acceptable, but the fact remains that when one can actually see shapes being formed and changed simply by the addition of varying frequencies of sound, it makes the formulae of the string theorist come alive.

The stated purpose of this chapter is to assist the reader in "connecting the dots" between several—perhaps many—varied disciplines so that the most difficult of the "transient ideas" that we must deal with are more easily overcome—primarily the idea of "relatedness." In conferring with Jana over the years as I was framing my thinking into more feasible thoughts which could then be woven together, it became increasingly apparent that the idea of "relatedness" is not an idea—it is a necessity!

The postulates that the cymaticist immediately gets at are astounding if one has never thought about them before:

- From formlessness comes form—just add sound.[58]

- Vibration underpins all matter in the universe. No matter can exist without sound and vibration.[59]

- Cymatics—The trigger for life? Spiritual traditions from many cultures speak of sound as having been responsible for the creation of life. The words of St. John's Gospel are a good example: *"In the beginning the Word already existed. The Word was with God and the Word was God."*[60]

---

58 John Stuart Reid

59 John Stuart Reid

60 My Note: This quote from the website translates "Word" as "sound." It will be my stated purpose to make it clear that the "Word" is the Incarnate "Word **OF** God—Jesus Christ."

It is important to state at this juncture that the mission statement of this remarkable group of scholars includes the position they take as a secular group, which is quite acceptable. But it will be my task to unite the exciting findings of this new emerging science with Biblical truth, as mentioned in footnote #60.

The original sound? Can you, in your sanctified imagination, even begin to comprehend the magnificence of an eternally imagined and created sound of Omnipotence as Yahweh uttered the very first notes of a song that would echo down through the eons of time and the inexhaustible halls of eternity and that would provide **all** that is, was and forever will be necessary to our life with Him? It is beyond the grasp of humanity to comprehend such a sound!

If you can find enough freedom of thought to first comprehend total silence and void—and then—an absolute explosion of gloriously beautiful and all-encompassing sound of such a caliber as to defy any and all other possible sounds. I can even imagine a sound with such magnificence and scope as to convey color, form and totally sweeping and overwhelming beauty that it does, indeed, defy definition and description. It is sound that is beyond sound, beauty that is beyond beauty, creative power that is beyond creative power. This is the voice of God, singing His creation into being—a voice never to be heard again until the annals of time have run out and the New Earth and New Heavens are put in place!

Scene II

DOES GOD'S SONG RELATE TO SCIENCE?

> *Consider this...*
>
> ## There is a "Song of God," and I believe that science has had its origins within the confines of the parameters set forth in that song.

Vibration underpins all matter in the universe.
No matter can exist without sound and vibration.

To see the periodic motions[61] that lie at the heart of matter is to lift the veils that conceal many mysteries of the universe.

Patterns generated by the CymaScope instrument often receive praise for their beauty, but beyond their obvious symmetrical perfection, what do they mean? Do they convey information?

The science of cymatics, the study of visible sound, is beginning to yield clues to one of the most challenging questions in science: **what triggered the creation of life on earth?**

---

61 Periodic motion, in physics, motion repeated in equal intervals of time. Periodic motion is performed, for example, by a rocking chair, a bouncing ball, a vibrating tuning fork, a swing in motion, the Earth in its orbit around the Sun, and a water wave. In each case the interval of time for a repetition, or cycle, of the motion is called a period, while the number of periods per unit time is called the frequency.

The trigger for life? Spiritual traditions from many cultures speak of sound as having been responsible for the creation of life. The words of St. John's Gospel are a good example:

"In the beginning the Word already existed. The Word was with God and the Word was God."[62]

String theorists made significant strides toward unraveling the great mysteries of the universe when they began to deal with quantum theory and the gulf which lay between the world of Einstein's theory of general relativity and the magnificent levels of activity that were exposed in the new and compelling equations generated by the necessities of the subatomic world.[63] The gulf between the world of general relativity and the world of quantum physics could be easily explained by the symphony conductor. And it is precisely here that the life of the **sanctified imagination** becomes exquisitely keen and actually becomes, in and of itself, a necessary key to understanding the role of sound and how it figures into the equation of creative energy.

---

62 *CYMASCOPE: Sound Made Visible*: Copyright 2008-2015: Cymascope.com: All rights reserved. These are direct quotes from several of the pages of a well-designed website dedicated to and describing the work of a team of scientists and scholars who are front-running the development of this new, emerging science. As a musician and author whose entire life's work has revolved around creation and the "beginnings" of all things beautiful, I am extremely excited about the work of these people. They are expanding upon ideas from hundreds of years back in scientific query and doing research which I believe will revolutionize our understanding of our universe and, yes, all creation. And their approach is of such a nature that their ideas can be empirically acceptable. Not only that, but the cymatists are able to back up their claims with visual and aural proof. When coupled with the claims of the string theorists, these new disclosures based on very old suppositions, will, I believe, be proven to convey information heretofore impossible to convey with authority.

63 Quantum mechanics (QM; also known as quantum physics, or quantum theory) is a fundamental branch of physics which describes physical phenomena at scales typical of the quantum realm of atomic and subatomic length scales, where the action is on the order of the Planck constant.

https://en.wikipedia.org/wiki/Quantum_mechanics

Act 2 — Scene II

Jana has provided me with the privilege of dialogue without critique while I sorted out some of these elusive realities which proliferate in my "non-scientific" mind.

Imagine: The musicians file onto the stage with a quiet dignity and seat themselves in their respective places—places of well-defined order—strings, reeds, brass, percussion—and intensely begin the process of tuning. Interesting. Interesting that they have to tune in the first place. Doubly interesting that this initial act of orientation and re-orientation to perfect harmonious sound is so elemental to producing musical beauty. Simply put—each and every one of these musicians' instruments must be accurately bound by just the right sonority, the perfect pitch level, the minute vibrations per second which will permit the creation of this "perfect sound." Ponder the multiplicity of single sounds piled one upon the other from all 80-odd instruments and the apparently chaotic and shapeless cacophony that results.

The wildly active tuning process will soon give way to the tightly organized harmonies of a magnificently written and just as magnificently re-played symphony or tone-poem. If the tuning were not at least approaching perfection, the entire experience of beauty and wonder would be derailed. It is when the master-musician, the conductor steps into his assigned place, takes his baton, raises his arms and arcs into being that first up-beat as precursor to that first down-beat that the miracle of creation in musical sound is allowed to come into being. It is only then that we are also allowed to roam the halls of wild fantasy or sublime peace and comfort or hear the cooing of a dove too beautiful to behold in reality. The miracle of music. What is it, really?

In pondering the progression of the apparently disorganized sound which defines the tuning process into the ultimate phenomenal, inex-

plicable marvel of such reason-defying beauty in sound as the orchestra launches into a Brahms symphony, one may be able to capture a brief glimpse of the necessity of quantum mechanics before one can grasp the beauty of what we see in our finite world. The frenzied gives framework to the organized. "From formlessness comes form. Just add sound."[64]

The next demand to be put on the sanctified imagination is to be able to harness the facts of the scientific world along with the massively commanding discoveries made recently and see how they fit into the emerging discoveries of the properties of vibrating energy as an essential tool or raw material in the creation process.

VIBRATION UNDERPINS ALL MATTER IN THE UNIVERSE. NO MATTER CAN EXIST WITHOUT SOUND & VIBRATION.

I believe that the researchers who are at the front line of the emerging science of cymatics are bringing some of the most profound and exciting discoveries of recent decades to our attention. One can trace their findings with great ease on YouTube or one of many websites dealing with their research in sound.

True, this is not new thought altogether. Even Leonardo di Vinci toyed with the idea that vibration had something to do with how things ended up being made. Ernst Chladni made great strides in the 18th century with his experiments on how vibrating sound could change shapes in the sand sprinkled on a copper plate when he bowed the plate with his cello bow. But the astounding new discoveries about vibration, frequency and sound have come about most recently because of the efforts of John Stuart Reid and his colleagues. Even though they make a clear declaration that they are a secular research group, they are providing a means for establishing

---

64 Cymascope.com John Stuart Reid: All rights reserved: 2008-2015.

experimentation which can be empirically documented, and which, by the way, fits so well into the evangelical framework of One God, One Creator and beautifully unified and organized patterns for raw materials available to create "something (everything) out of nothing." **Nothing, that is, but vibrating sound.**[65]

Before there was anything, there was void. Neither the Scriptures nor any other ancient Holy Writings gives us a clear definition of "void." It is simply stated that *"the earth was without form and **void"*** in the Book of Genesis.

The concept of void and nothingness is evasive. Again, you see the need for sanctified imagination and a counterpart of your own persona with whom to dialogue. The Scriptures and Holy Writings of the ancient Hebrews are vague but consistent in their description of the character of the universe or world or whatever it was before the dawn of the creation event. Imagine the depth of the silence in that void. The next event was simply that "God said" (sang).

I like to ask the question, "What was the very first sound ever heard?" Since most people have never given thought to that particular question, it has the potential of inspiring a vast array of responses. Thoughts that have not been entertained before can come springing to the surface to ignite a forest fire of ideas. This is the way that Jana and I dialogue, and this is the way that I learn so much that is not written in my books.

Whatever direction your mind as the reader takes, consider this: There was void and nothingness. There was a sound. It was the sound of the voice of God. It was a word, or it was a song. But it was the first sound—vibrating energy. And if you can allow your mind to embrace the

65 https://www.youtube.com/watch?v=KU84ckD1AcA.

all-powerful and eternal capacity of the voice of Almighty God, you can begin to give place to the massive energy output that would ensue. All of whatever space or darkness there was at that pivotal moment would be set to moving and vibrating and pulsating.[66]

"Into the great voids of space came a sound and matter took shape."[67]

TO SEE THE PERIODIC MOTIONS THAT LIE AT THE HEART OF MATTER IS TO LIFT THE VEILS THAT CONCEAL MANY MYSTERIES OF THE UNIVERSE.

Historically, the pursuit of knowledge has a story of its own to tell. Scholars have, since the beginning of recorded time, been seeking answers to the same basic questions about origins: the origin of the universe; the origin of man; the origin of language; the origin of music; of medicine; of art; of philosophy. A related question is always in the dimly-lit wings: Why? Why are we here in the first place? What—really—is our relevance as the only creation with the ability to think cognitively, and, above all, to express what we think? Is there a purpose to our presence here and are we alone?"

In order to understand and answer that question intelligently, we must have a deeper understanding of the universe to begin with. The conductor who sets out to open the beauty and secrets of a great oratorio or symphony must first get into the music, dissect it and analyze it in that attempt to see inside the mind of the composer and thereby to ascertain that composer's intent in writing it in the first place. To be **there** at the moment that the composer puts pen to paper. Because we are, as that conductor,

---

66 Personal Footnote: I recognize that my evangelical world view enables me to make this connection very easily, and I, in no way, seek to impose that view. I still hold the opinion that to have a differing world view in no way negates the validity and worth of scholarly research which holds to another view.

67 Peter Guy Manners, M.D.

only mirroring the composer's **intent,** it is immensely imperative to **know** and to **understand.**

Robert Shaw, the great choral conductor of the 20th century, was a master of this concept and approach to his art—choral conducting. In later life, he finally could bring this incredibly insightful approach to the symphonic orchestra and literature as well. He was truly an icon in the world of music, but I do not believe that his only contribution to mankind was exclusively musical. I believe that one of the prime purposes of his sojourn on this planet was to teach, at least some of us, this irrefutable lesson: that you cannot know the intent of the Creator without knowing how He divined from the **very** beginning to facilitate that intent. Of course, his venue for teaching this profound lesson was not theological or even academic: it **was** musical. This was his worship. And why not? Since it is my convinced opinion that musical resources, vibration, frequency, pitch, along with their cousins, harmony, rhythm and form, are the absolute raw materials of the creative chronicle.

In his uniquely insightful biography of Shaw, Keith C. Burris, makes this so clear:

> "Shaw's (preparatory) editing involved some direction for virtually every note, in every musician's part—instrumentalists, chorus, and vocal soloists ... In preparation for conducting the Bach *Mass in B Minor* at the University of Kansas in 1990, Shaw sent the UK chorusmaster **three** piano-vocal scores he had edited. The first 'clarified choral and concertist roles—the second dealt with tempi—and a third with phrasing—attention to the small cell....'"[68]

---

68   *Deep River: The Life and Music of Robert Shaw*: Keith C. Burris: GIA Publications, Inc., 7404 S. Mason Avenue, Chicago, IL 60638: www.giamusic.com: 2013.

It may be difficult to make the leap of intellectual faith to see the corollaries between huge musical scores and their performance, and the creation chronicle also as a huge "musical" event. Unless, of course, two concomitant things take place: 1) the reader allows his or her sanctified imagination to come into full play, and 2) the reader finds that rare, intimate friend, the alter-ego and allows at least some degree of dialogue with it. Remember C.S. Lewis' deep respect for his imagination. His definition of imagination was powerful: "Reason is the natural organ of truth, but imagination is the organ of meaning. Imagination ... is not the cause of truth, but its condition."

Michael Ward,[69] English scholar, wrote about Lewis:

> "In Lewis's view," according to Ward, "reason could only operate if it was first supplied with materials to reason about, and it was imagination's task to supply those materials."

The next piece to this growing puzzle is the gradual accruement of the knowledge of our universe and its almost unknowable footprint and how it came into being. Human knowledge has grown slowly and incessantly over thousands of years, and has been gradually opened up to humankind as God has seen fit. *The secret things belong to the Lord God, but those things which are revealed belong to us and to our children forever.*[70]

James Burke, best-selling author and television producer for PBS and BBC-TV, said,

69   Michael Ward is a Senior Research Fellow at Blackfriars Hall in the University of Oxford, author of *Planet Narnia: The Seven Heavens in the Imagination of C.S. Lewis* (OUP, 2008), co-editor of *The Cambridge Companion to C.S. Lewis* (CUP, 2010), and presenter of the BBC television documentary, *The Narnia Code* (2009).

70   Deuteronomy 29:29: NKJV: 1979.

Act 2 — Scene II

"We are what we know—and as the body of knowledge changes—we also change."[71]

When Galileo invented the telescope, the concept of truth changed. Truth did not change, but the concept of truth did. Prior to his cataclysmic discovery of the telescope, it was considered truth that the world was flat. And that **it**, the earth, was the center of the universe. The changes that were precipitated upon scholars by his telescope effected architecture, music, literature, science, politics **and** religion. A world revolution of sorts followed as the impact of the new view of the universe expanded.

The advent of discoveries such as the telescope and the microscope have revolutionized the way in which man sees his world and all that is in and around it. **Imagine**, if we can perfect that science of making sound visible, if we can track **how sound changes the forms that it creates**, if we can begin to relate how this compares to the way in which beautiful music emanates from the formless noise of a tuning orchestra and if we can then understand a bit more clearly how that music is really made from only 13 tones—tones which are so vitally altered by one or two vibrations per second difference in that tuning—once more mankind would see his view of the world, the universe, and all creation being turned upside down. This new revelation of visual sound would be, in my mind, the most transcendental of all new revelations because it would provide an **explanation** for **all other** discoveries and revolutions of thought.

The *cymascope* is a new type of scientific instrument that makes sound visible. It is capable of transmitting frequencies of myriad pitches, volumes and timbres onto or into a chosen medium. As the frequencies

---

71   James Burke: *The Day the Universe Changed*: Little, Brown and Company: September 1986.

91

change, the shapes on or in the medium also change.[72] The science is so new that present mathematical techniques cannot predict what pattern will form on the membrane or medium. But the very fact that this instrument is able to demonstrate the facility of sound to create and the necessity for the presence of sound **to shape and create** is phenomenal.

In synopsis, here is a snapshot look at the new science, cymatics:

"The generic term for the patterns of vibration that occur on the surface of an object when excited by an incident sound is 'modal phenomena,' a field of study that covers everything from vibrations in suspension bridges, to vibrations in body parts of cars, to the effects of sound on the human skeleton and internal organs. In the 1970's this branch of science was named 'cymatics' by Swiss doctor Hans Jenny, a word that derives from the Greek 'kyma,' meaning 'wave' and the inspiration for the name of our CymaScope instrument. The classical view of modal phenomena is that modal patterns form as a consequence of the natural resonant frequencies, or modes, of the object or membrane; current mathematical techniques used to describe this class of phenomena say nothing about the quality of the exciting sound. Musical sounds contain many harmonics so when a circular membrane is excited by a complex musical sound the resulting modal pattern(s) are, naturally, also complex.

"When the microscope and telescope were invented they opened vistas on realms that were not even sus-

---

72   http://cymascope.com/cyma_research/index.html.

pected to exist. The fields of biology and cosmology would have remained closed to us without these instruments. ...The CymaScope holds the same potential for advancement as the microscope and telescope and its applications are on the brink of touching every aspect of human endeavor."[73]

PATTERNS GENERATED BY THE CYMASCOPE INSTRUMENT OFTEN RECEIVE PRAISE FOR THEIR BEAUTY, BUT BEYOND THEIR OBVIOUS SYMMETRICAL PERFECTION, WHAT DO THEY MEAN? DO THEY CONVEY INFORMATION?

More than forty years ago, the first vestiges of curiosity about the universe came into my mind; not that I was as curious about the universe itself as I was about the vast evidences of musical presence that were obvious to me **in** the universe. I was fascinated by the notions that music could heal, that an infant's IQ could be enhanced by the music it was exposed to, that everything in the universe had a harmonic vibration which often could be heard by the naked ear. I was amazed that Beverly Sills could shatter a crystal glass with harmonics just by singing a "high C." When I found that the Aurora Borealis, one of my very favorite natural phenomena, had a voice and could and **did**, in fact, sing, my excitement was all but uncontainable. It was light waves, sound waves, radio waves and electro-magnetic activity all rolled into one magnificent aura of ethereal beauty! And all of these marvelous manifestations had musical implications, if not out and out musical connotation.

---

73   http://cymascope.com/cyma_research/index.html.

It was incredibly interesting for me to discover that science was beginning to say that light waves, sound waves, and mathematical ratios were comparable entities and could be compared one to the other. The vast activity of the electrons and atoms could be seen and measured at a subatomic level, and their activity was mathematically related to the ratios of light and sound waves and harmonic proclivities. It was mostly an intuitive acknowledgement and the research up to that point was not adequate enough to support a well-constituted theory. It was beyond my comprehension: were they saying that there is a rhythmic/musical component to our universe that they had not yet exposed and that they could not really empirically document? What slowly dawned into my awareness was that everything in the universe is somehow connected and the connection, in all likelihood, was to be recognized as being musical in nature. But where was I to look for reinforcement of my sacred imaginings?

And so, for the next four decades I pursued my shadow of knowing, vaporous as it was, and over the years my thinking began to take shape. Yes, everything is connected. And I discovered that Albert Einstein thought so, too, and it was that shadowy idea of the reality of the "unified theory" that drove him to keep seeking. Then I discovered that in the deepest halls of scientific discovery there was, indeed, a growing and somewhat substantive amount of research leading to the acceptance of these facts: 1) there is much more that we do not know than there is that we do know, 2) it is possible that this thing called vibration or frequency seems to be consistently evident in many places in many ways, 3) there is an increasing amount of common ground that will give us an understanding of and explanation for the great contradictions that are so common in scientific research—like the gulf between general relativity and quantum physics. And the list of problematic issues goes on and on. The increase of understanding that came with the sudden growth of string theory in 1984 made great strides in solving some of these issues, but did not quite make the grade toward full accomplishment.

However, we must acknowledge that scientific achievement has historically been the doorway to new information and a new growth in knowledge—such as the development of the telescope and the microscope, both of which opened entirely revolutionary views and attitudes concerning our universe and all that is in it.

Does the imagery conveyed by the cymagraphs portrayed in the laboratories of the cymatic physicists convey information? Mostly assuredly, and it is becoming more and more evident that this could be another of those revolutionary instances when a man's curiosity and discovery changes the entire direction of man's thinking about his universe.

THE SCIENCE OF CYMATICS, THE STUDY OF VISIBLE SOUND, IS BEGINNING TO YIELD CLUES TO ONE OF THE MOST CHALLENGING QUESTIONS IN SCIENCE: WHAT TRIGGERED THE CREATION OF LIFE ON EARTH?

And now in the 21st century another incredibly revolutionary idea is burgeoning: the idea that without sound there would be nothingness. That it is sound that holds the entire universe together after all. That emerging science is called cymatics, and was so named by Hans Jenny, Swiss medical doctor and scientist.[74]

Cymatics is defined as: "the study of visible sound co-vibration, a subset of modal phenomena. Typically the surface of a plate, diaphragm,

---

74 Hans Jenny (b 1904 d 1972) was a Swiss medical doctor and scientist who studied visual sound intensively. Jenny published his first volume, *Kymatic*, in 1967 and his second in 1972, the year he died. Jenny coined the word *Kymatik* ('cymatic' in English) from the Greek 'Kuma' meaning 'billow' or 'wave,' to describe the periodic effects that sound and vibration has on matter.

or membrane is vibrated, and regions maximum and minimum displacement are made visible in a thin coating of particles, paste, or liquid."[75]

The study of string theory has already laid the foundation for the hypothesis that matter is finally built of nothing more or less than vibrating energy. In my layman's mind, though, I can see how a very real problem surfaces: there is no vehicle for proving this except for the mathematical equations that do, in fact, support the supposition. Equations are not considered empirical proof; it is still theoretical.

In 1997, John Stuart Reid, an acoustics engineer from the UK, traveled to Egypt to research the acoustical properties of the King's Chamber in the Great Pyramid. His findings were phenomenal and very suggestive. There was, besides what he intended to see, an underlying item of interest. He underscored the suspicion that vibration and the development of language could be intertwined, thus strengthening the idea that **all things in the universe** are connected or related, one to another:

> "Reid published his research results in Egyptian Sonics containing photographs of the cymatic patterns that formed on a PVC membrane he stretched over the sarcophagus. The experiment was designed to study the resonant behaviour of the granite from which the sarcophagus is fashioned. Since many of the images strongly resemble ancient Egyptian hieroglyphs Reid postulated that the inherent resonances of granite, when radiated as a com-

---

75 Cymatics (from Greek: κῦμα "wave") is the study of visible sound co-vibration, a subset of modal phenomena. Typically the surface of a plate, diaphragm, or membrane is vibrated, and regions of maximum and minimum displacement are made visible in a thin coating of particles, paste, or liquid. Different patterns emerge in the excitatory medium depending on the geometry of the plate and the driving frequency. https://en.wikipedia.org/wiki/Cymatics.

plex sound field during crafting of the stone, might have influenced the development of hieroglyphic writing. Reid subsequently began experimenting with instrumentation that would enable an accurate visual equivalent of sound to be created from any audible sound."[76]

Out of this experience came the invention of what I consider to be one of the most dramatic and important inventions of the 20th and 21st centuries, the cymascope. It has been held that the sub-microscopic particles of matter behave in very well articulated and consistent ways. There is no happenstance in Mother Nature's constructs. Whether one is dealing with gas, liquid, solid or plasma, those atoms and/or molecules move and interact **in perfect harmony**. That, we have known since pre-atomic age. Even the orbital paths within the atoms and molecules are well-disciplined particles and are consistent in their behaviors.

The cymascope, even in its present developing state, has made it possible to see, to track and to photograph those behaviors and the patterns of the paths they create. There is a mind-rattling and reason-defying perfection in these patterns: they are uncannily similar and, more than that, they make it clear that matter is actually geometry in motion. The once hidden realm of sound is suddenly opened to us and provides science with a window into the very heart of sonic vibrations. By mapping where the orbital paths of atomic and sub-atomic particles cross and by connecting those points of intersection, Euclidean geometry becomes apparent: we can see virtual geometric structures interacting with space. Incredible!

76 http://www.cymascope.com/cyma_research/history.html.

But only by the engagement of your sacred imagination can you actually grasp the importance of these new ideas. This quality of free envisioning is what fires the furnace of great discovery. The question arises quite naturally: "What triggered the creation of life on earth?" With the growing body of provable evidence that it is, indeed, vibration of energy—sound—that actually makes and changes forms and shapes, that question becomes more and more pressing. It begs resolution. The short answer is just that: **sound.** But the inadequacy of that answer is huge and unforgiving. There must be an extension to that question: **What sound and from whence did it come?"**

THE TRIGGER FOR LIFE? SPIRITUAL TRADITIONS FROM MANY CULTURES SPEAK OF SOUND AS HAVING BEEN RESPONSIBLE FOR THE CREATION OF LIFE.

The words of St. John's Gospel are a good example: *"In the beginning the Word already existed. The Word was with God and the Word was God."*

The cymatic scientists and scholars who are at the heart of this emerging and exciting new science are making no claims of being at all evangelical in their stated philosophy and mission, but they certainly do recognize some of the salient properties of the "religious" approach to their research: 1) they are aware of and give homage to the spiritual traditions, both contemporary and ancient, of other cultures around the world, 2) one of their driving quests is to isolate and learn to effectively use the healing properties of their discoveries because they do believe in that facet of their work, 3) they are encouraging to other scholars who are interested in their progress and urge participation regardless of the spiritual orientation of those people.

This is a direct quote from one of their website pages:

> *"In the beginning the Word already existed. The Word was with God and the Word was 'sound.'"* ('Word' meaning 'sound.')[77]

On their website, Reid and his colleagues quote, *"And the Word was God"* as *"the Word was sound."*

I do not know where the concept of the word "Word" being defined as "sound" originated, but the Greek transliteration in the Book of St. John, quoted as the source of this sound byte of wisdom, is:

1) a word, uttered by a living voice, embodies a conception or idea

2) what someone has said

   a) a word

   b) the sayings of God[78]

The theology of those verses from St. John, named for himself and as penned by him, brother of James, son of Zebedee and early follower of John the Baptist and Jesus Himself, is that this "Word" was the activating force of creation and all life, incarnate in the Person of Jesus of Nazareth, Jesus Christ. Further, it is stated several times in the New Testament that He was present with God before creation, and by Him was everything made that "was made." There seems to be no inference in the Greek lan-

---

77 http://cymascope.com/cyma_research/biology.html.

78 https://www.blueletterbible.org/lang/lexicon/lexicon.cfm?Strongs=G3056&t=NKJV:John !:!

guage that "Word" could or should mean simple sound. From whence that "sound" and to what end without that divine necessity?

This theology will be further explored in ACT III, "Faith and the Theology Behind That Faith." It is true, however, that many cultures entertain metaphors and myths about the creation chronicle which include strong references to "sound" as a tool, even a driving force of creation, and especially sound as a musical sound—singing. Again, I refer to Narnia and *The Magician's Nephew* where the singing lion, Aslan, who is the counterpart or metaphor for Jesus, the Lion of God sang into being the Land of Narnia. And, of course, the singing creator god, Iluvitar, in the tale of Ainulindale in the *Silmarillion* of J.R.R. Tolkien, and how he thought the Ainur into being with his songs.

I think the idea of the singing creator had to come from somewhere, and because I also believe that the creative mind is but a dim, smoky reflection of the infinite and all-pervasive creativity of God the Master Creator, it follows that all of these ideas, metaphors and analogies should be parallel in some manner or other. One Master Creator—one master plan.

Words come easily to us human beings for the most part. We can go on and on about the things and ideas that most completely interest us. But after spilling much ink, and printing many pages, we can still be left with the same unanswered questions: 1) How was this universe put into being? 2) What triggered the life that this universe supports and nurtures? 3) Why was this universe put into being?

Many years ago, I restated that *"God created the heavens and the earth"* because it is written in Holy Scripture,[79] and I accept that as final

---

79 *God's Song* 1st edition: Liberty University Press: Lynchburg VA: 2010; *God's Song*, revised 2nd edition, Publishers Solution: Forest, VA: 2017. This book is the "pre-quel" to *God's Opera*.

truth. I feel no inclination to defend that belief—it simply **is** for me. There is too much peripheral evidence to allow me to disregard this belief as legitimate and viable—evidence in literature, in music, in legend, and, yes, even in science. From a personal and, maybe, even a literary point of view, it is satisfying to be able to start there, but it certainly does not answer the question stated above for everyone, and especially for those in the academic and scientific worlds.

And God is not openly forthcoming in His portrayal of the creation methods. He only gives us enough information to allow faith to survive: *"And God said—and it was."*[80] There is, however, enough information in that statement alone to verify the idea that it could be, indeed, from a scientific viewpoint, **sound** which was the bottom-line force that made the creation of the universe and all that is in it. And it was that **sound,** whatever that sound was, that triggered the formation of structure and shape, and the myriad variations of that **sound** that made possible everything that we know of as matter. This much, the cymaticists are comfortable in stating: **"Everything owes its existence to sound … Sound is the basis of form and shape."**[81]

It has almost become a matter of intellectual integrity to preserve one's conviction that science and art, or science and theology must be at odds one way or another. Is this fact or fallacy? Or, maybe, it is simply that over the generations of discovery, mankind has become comfortable with the trends of thought that support the idea that certain disciplines are bound to be at odds. The free thought engendered by imaginative encounters with one's own inner intellect, carefully disciplined, might find boundless new territories acceptable to dispel the dichotomies.

80   Genesis 1: NKJV, Nelson Thomas, Inc., 1982.

81   http://www.cymascope.com/cyma_research/history.html.

Allowed to be inspired by one's sacred imagination, the beauty and excitement that can result from beginning to see this integration of science and beauty (art, if you please) as wholesome would be, as I see it, an awesome stride toward real understanding and resolution. And given today's new and emerging thoughts and discoveries in both fields of endeavor, I am encouraged to expect this integration to happen.

I believe there is a "Song of God," and I also believe that it is entirely possible that science has, in fact, had its origins within the confines of the parameters set forth in that song—consider and confer with your innermost self—activate your sacred imagination.

$\mathscr{S}cene$ *III*

GOD'S SONG MAY BE SCIENCE INCARNATE

> Consider this…
>
> *I believe that sound—vibration—is the force that initially caused particles to band together into a vast multitude of differing shapes and angles, governed by the frequency of that sound.*

Energy = Mass x Speed of Light
E = MC2
The Energy of Motion Turns into Mass
and
SOUND HOLDS IT TOGETHER!

The Universe is made up of:
Twelve Particles of Matter
Four Forces of Nature

THE INTRIGUING IDEA OF ENERGY & SOUND AS THE GENESIS OF THE UNIVERSE.

One of the most intriguing observations that I have made over the last forty years of research into creation and "beginnings" is the remarkable way in which science has come so close to truth as it is por-

trayed in Scripture. Those of us who share the evangelical world view as the foundation of our grasp of the universe and all that is in it have a very different understanding of scientific accomplishment and discovery. Scientists are to be commended for the tenacity with which they have pursued solving the mysteries of creation and life on this planet. More often than not they come extraordinarily close to the defining verities, but because truth cannot be definitely proven and cannot be empirically documented most of the time, they are hampered by their training in accepting the connections that may seem quite basic to the spiritually conservative layman. And so it was with Einstein, who spent the last thirty years of his life writing so elegantly about the mind of God. At a dinner party in Germany in 1929, when asked if he were religious, he was heard to say:

> "Yes, you can call it that. Try and penetrate with our limited means the secrets of nature, and you will find that, behind all the discernible laws and connections, there remains something subtle, intangible and inexplicable. Veneration for this force beyond anything that we can comprehend is my religion. To that extent, I am, in fact, religious."

Shortly thereafter, he was interviewed by the poet and propagandist, George Sylvester Viereck, who taunted him with the following exchange:

Q: "To what extent are you influenced by Christianity?"

A: "As a child I received instruction both in the Bible and in the Talmud. I am a Jew, but I am enthralled by the luminous figure of the Nazarene."

Q: "You accept the historical existence of Jesus?"

A: "Unquestionably! No one can read the Gospels without feeling the actual presence of Jesus. **His personality pulsates in every word. No myth is filled with such life.**"[82]

There are so many things in this universe of ours to intrigue us. One of those things is language. Obviously, when Einstein became engaged with any thought or idea, he also became deeply **involved** with that thought or idea. It was that quality that defined his incredible intelligence. But in the first half of the 20th century so little was known about the very things he was grappling with. Yet his choice of **words** in answering Viereck's carefully designed questions is still, in the first half of the 21st century, one-hundred years later, graphic to the point of being prophetic:

"His (Jesus') personality **pulsates** in every **word**. No myth is filled with such **life**."

If we take seriously (and I do) the findings of this newly emerging science, cymatics, we must admit that his choice of words was enigmatic, if not profoundly revealing, to say the least. To "pulsate" is to vibrate. "Words" are, in fact, vibrating energy, as is "song." So the vibration in "every word" of the Nazarene's becomes worthy of meaningful consideration. And, by inference, those words become the quality that gives life—therefore, no myth **could be "filled with such life."** Only the truth conveyed in the words of the Nazarene, by that line of thought, should be understood as the underpinning of the creation chronicle.

---

82 Walter Isaacson, the CEO of the Aspen Institute, has been chairman of CNN and the managing editor of Time magazine. His new book, "Einstein: His Life and Universe," was published May 13, 2008. http://www.faithstreet.com/onfaith/2007/04/27/einstein-and-the-mind-of-god/3763.

It takes an open mind and an imagination, also open full-wide and which is dedicated to serious introspection about our capabilities, to recognize, to internalize and to develop really new thought. "Is this idea worth my consideration and energy to pursue?" The idea that I am referring to is simply this: science is now saying that it is **sound** that is not only the cohesive factor in our universe(s), but **sound** is also the inevitability that produces form of any and all kinds. The new emerging science of cymatics has recently made this bold statement: "Without sound there would be no matter."

The fascinating thing about all this is that this new and emerging discipline, cymatics, has a means of eventually being able to empirically document its ideas as viable scientific research. When one can enter the laboratory of cymatic research and see before one's very eyes the miracle of form coming out of formlessness, it is a astounding experience. The idea of sound, vibration or pitch being used as the tools and the raw materials of creation is not a new idea. In fact, the embryonic idea has been a significant part of scientific research and mathematic rumination for centuries. It has also been a source of irresolution and conflicting theories. Cymatics as a science, in my opinion, has the capacity to become a revolutionary source of undeniable discovery, and it commands attention.

> "Into the voids of space came a sound and matter took shape."[83]

> "If you want to find the secrets of the universe, think in terms of energy, frequency and vibration."[84]

83   Peter Guy Manners, MD: British Osteopath, qualified in England and Germany, in Natural and Electromagnetic Medicine: Copyright Cyma Worldwide. All rights reserved: https://www.youtube.com/watch?v=_Y4supY0PWw

84   Nikola Tesla, 1856-1943: Serbian American inventor, electrical engineer, mechanical engineer, physicist, and futurist best known for his contributions to the design of the modern alternating current (AC) electricity supply system.

TWELVE PARTICLES OF MATTER: FOUR FORCES OF NATURE

Early on in the history of mankind, scholars began to think deeply about their world and the universe of which it seemed to be a part. They were immediately faced with the necessity of finding reasons for what they observed. It became evident that there had to be some connectivity in the shapes and forms, the sounds and rhythms around them which were so easily observed but which remained such hidden and enigmatic issues. There was mathematics. There was art—always there was art—everywhere—artistic shapes in nature—artistic sounds in birdsong, the wind, the babbling brook. The things to be considered were myriad. The stars which they could not even comprehend, much less number or describe. And on and on—"ad libitum."

Soon they began to formulate well-defined subjects of study. And that very discovery of different fields of naturally delineated study set off a new quest—the similarities began to manifest themselves in spite of themselves. Why? Where did those similarities come from? What did they mean, if anything?

They, the scholars and thinkers, tried to relate musical frequency to the mathematical equations which were surfacing in serious universities. They tried to answer questions about the earth and heavens based simply upon what they could observe with the naked eye. There seems to have been a persistent drive to relate the laws of nature and the observed universe and certainly there was a persistent drive to understand how these laws worked and where they came from in the first place. It was not long before they began to see the necessity for instruments to help them in their quests for knowledge.

We have already looked at the tremendously culture-altering impact of the telescope and the microscope. Science was born. Time passed and

tremendous advances in knowledge became extant and were made largely available to the common man. But one thing evaded the most dedicated and most intelligent—the secrets of the universe—its creation—its sustenance—and the reasons for its having come into being in the first place. The drive to understand our own being and the existence of our world and universe is likely the most important driving force behind the evolving of all knowledge. And that became the pathway for cultural development as well, because, as James Burke wrote: "...We are what we know—and as the body of knowledge changes—we also change."[85]

And so it has been all down through the halls of time: **query spawns discovery; discovery spawns knowledge; knowledge spawns more queries.**

And so we arrive at the 21st century: the age of knowledge and information. More knowledge and more information than man can assimilate. The age of specializing. And the encroaching ramifications of an age which also spawned a cultural **trend** that positioned man as the ultimate authority and the center of all things known and sought after.

> "The Age of Enlightenment or simply the Enlightenment or Age of Reason is an era from the 1620s to the 1780s in which cultural and intellectual forces in Western Europe emphasized reason, analysis, and individualism rather than traditional lines of authority—It challenged the authority of institutions that were deeply rooted in society, especially the Roman Catholic Church; there was much talk of ways to reform society with toleration, science and skepticism."[86]

---

85   James Burke: *The Day the Universe Changed*: Little, Brown and Company: September 1986.
86   https://en.wikipedia.org/wiki/Age_of_Enlightenment.

Act 2 — Scene III

Wikipedia, the free encyclopedia is not, in my mind, an ultimate authority to be overused. However, even though it does not carry the credentials of *Encyclopedia Britannica*, it does have a place in our world and it does contain a wealth of readily accessible information. In this case, I feel free to quote its definition of the Enlightenment as credible.

The Age of Reason, or the Enlightenment, was a movement of monumental scope. The entire landscape of socio-economic and academic culture was changed over time as a result of this sweeping revolution. No longer was the autocratic power of the church, in particular the Catholic Church, accepted as the authority through which came all delegation of and development of station or class and knowledge. The bourgeoisie began to demand education and a voice in matters, and most amazingly, they finally got it. The church began to recede as the "be-all" and "end-all." Man wanted his own freedom, and he wanted to be able to define that freedom for himself. The intellectual liberation of the masses in the 1600's was a good thing, and the resulting changes in lifestyle were necessary and magnificent. The Age of Reason was, perhaps, solely responsible for initiating the progress that we of the 21st century now see as being normal—without that intellectual breakthrough, the Industrial Age, the Space Age, the Technological Age and the myriad other advancements may never have become reality.

On the other hand, I wish to challenge the reader with this idea: perhaps, what we call man's invention may be nothing more than the discovery of those things that God Himself chose to grant to His creatures and which He, in His omnipotent power had already created and put into motion for us to discover. By that I mean that, at the moment of ultimate creation, God had already designed and put into effect every law, every

boundary, every component part for ultimate creation and the control of it. That eternal design was already perfect in every aspect.

Then, as time progressed, Pythagoras discovered the Golden Mean; Fibonacci discovered the perfect sequence of numbers; Isaac Newton discovered his Law of Universal Gravitation; Einstein discovered the General Theory of Relativity; Michio Kaku discovered the expansion of that theory into String Theory; and John Stuart Reid discovered a way to demonstrate modal vibrational phenomena. To say nothing about the progress of engineering feats: the Wright brothers take-off into the air, the development of electricity, the movie camera, the jet propelled engine, computers, cell phones, GPS—

Here is the astounding and commanding realization: all of these things have a connection. There is an undeniable connectivity among all the immeasurable facets of nature, medicine, art, science, mathematics, things too myriad to even enumerate.

One of the astounding "discoveries" of the scientific world is the discovery of the twelve particles of matter and the four forces of nature. **Particle physics,**[87] that branch of science which deals with matter and its genesis, is so infinitely complex that to even attempt to explain it would be hugely complicated and confusing. In addition, modern thinking con-

---

87 Particle physics is the branch of physics that studies the nature of the particles that constitute matter (particles with mass) and radiation (massless particles). Although the word "particle" can refer to various types of very small objects (e.g. protons, gas particles, or even household dust), "particle physics" usually investigates the irreducibly smallest detectable particles and the irreducibly fundamental force fields necessary to explain them. By our current understanding, these elementary particles are excitations of the quantum fields that also govern their interactions. The currently dominant theory explaining these fundamental particles and fields, along with their dynamics, is called the Standard Model. Thus, modern particle physics generally investigates the Standard Model and its various possible extensions, e.g. to the newest "known" particle, the Higgs boson, or even to the oldest known force field, gravity.[1][2]

tends that there are four forces (interactions) of nature[88] [89]that can be helpful in explaining these twelve particles and their behaviors.

Even a minute understanding of logic could then allow one to deduce: If all things are composed of the same elemental, infinitesimal particles and interactivities (forces) as understood by science, then all things would be bound to be related and, in some way or another, to interact as members of the same creation.

The next question, of course, is Who the Designer of these phenomena could be and what genus of raw materials or elements did He use with which to accomplish His creation?

There are postulations afoot today that seem to accept the idea that there was a void, there was a time when something cataclysmic occurred and that the indications are that form did, indeed, come out of formlessness. There is, also, a new and growing expression of the fact that sound—vibration—could be that force that initially caused the particles to band together into a vast multitude of differing shapes and angles, governed by the frequency of that sound, much as we can observe from listening to a symphony orchestra tune.

For the sake of simplification and organization of thought, imagine, in your sanctified imagination, that there is a Supreme Being Who has the capacity for song, and that He initiated that song abruptly

---

88 According to the present understanding, there are four fundamental interactions or forces: gravitation, electromagnetism, the weak interaction, and the strong interaction.

89 Fundamental interactions, also known as fundamental forces or interactive forces, are the interactions in physical systems that don't appear to be reducible to more basic interactions. There are four conventionally accepted fundamental interactions—gravitational, electromagnetic, strong nuclear, and weak nuclear. Each one is understood as the dynamics of a field. The gravitational force is modeled as a continuous classical field. The other three are modeled as a discrete quantum field, and exhibit a measurable unit or elementary particle.

and that the energy and quality of that song was the creating force of all that we know.

Could the quality and quantity of energy expended and set into motion at the moment of original creation—the moment of the Word being spoken or sung for the very first time—account for the ineffable and incontrovertible nature of energy as we understand it today? Is that possible? Consider…

Scene IV

FROM FORMLESSNESS COMES FORM—JUST ADD SOUND

Consider this…
The universe was created … one way or another,
it was created.

Random thoughts trigger cohesive thoughts.
-Jana

Sound takes random particles and shapes them into forms.
Sound is energy we hear made by things that vibrate.
-Chris Woodford

Matter is geometry in motion.
From Cymatics: Geometry
-Cymatics.com

Everything is sound. In the beginning was the Word—and the Word is sound.
From formlessness comes form—just add sound.
-John Stuart Reid

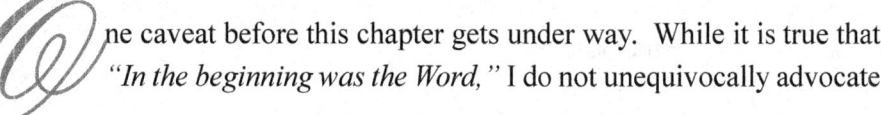ne caveat before this chapter gets under way. While it is true that *"In the beginning was the Word,"* I do not unequivocally advocate

the idea that to say that the "Word is sound" is true. What our sacred writings actually say is: *"In the beginning was the Word and the Word was with God and the Word **was God**."*[90] But once again I am constrained to repeat that this work is not in any way intended to be an apologetic for theological ideas. I am simply reiterating the point from which I define my own personal beliefs and upon which I build my theories.

The implications are incredible and have already been discussed in ACT I. Further resonating thoughts, if considered carefully and with the engagement of one's sanctified imagination—one's personal Jana, if you please—the idea is, unreservedly, that the first resounding sound ever heard in all of time and eternity was made by the singing voice of the living Word of God—Jesus—as God the Father's singly appointed Creative Force. Even though this book has no interest as a theological apologetic, there is a fascinating consideration in the New Testament which should be included in the body of ideas being explored. In describing the Personhood of Jesus, the Book of Colossians states clearly, *"By Him all things were created and by Him all things consist."*[91] There is a vague acceptance of the fairly new idea that sound is the cohesive force in all of creation. This being the case, the idea of the **"Word of God"** (inferred as the "Song of God" in this writing) should be examined as the possible source of that "sound," even without going any further into the Messiah-ship of Jesus in this context.

But the primary goal of this book is not to foster an exegesis of this or any other text. The goal is, simply stated, to present a fresh new glimpse of the marvel of creation and Creator, and to give adequate credibility to the concepts presented, in the hope that other scholars, as mentioned before, will pick up the banner and further the research begun here.

90   St. John 1:1-3, New Testament, NKJV.
91   Colossians 1:15-18.

Act 2 — Scene IV

RANDOM THOUGHTS TRIGGER COHESIVE THOUGHTS

We have arrived at the near-end of ACT II in this chronicle of the magnificent beginnings of all things, and if one unmanageable thought has emerged, it could easily be that there is too much information, too much knowledge, too much variation, too much subject matter to be dealt with easily and once for all. Besides trying to deal with and codify so much broad knowledge, there is the indomitable fact that all of the things we are struggling with are changing. Go back to the statement made by James Burke in his powerful treatise on *The Day the Universe Changed*:

> "We are what we know—and as the body of knowledge changes—we also change."[92]

And so, we are faced with the nearly insurmountable task of doing the same thing that shapes the idea of "form out of formlessness"—so much data to deal with creates a monumental task—the task of relating one thing to another continues to expand and increase. And that may be analogical in itself. The universe was created—one way or the other, it **was created**. For hundreds of years that was viewed as fact, not understood, but still it was viewed as fact. In recent times, the quest for knowledge and answers has burgeoned. The findings of Edwin Hubble in 1929 only added to the quandary—what he saw suggested that the far-distant galaxies were moving away from us. Did that mean that they were expanding?

The advent of the Hubble telescope has made it more feasible to believe that, indeed, the universe is expanding, but this remains a point of disagreement in many scientific circles. However, that idea is consistent

---

92   James Burke: *The Day the Universe Changed*: Little, Brown and Company: September 1986.

115

with the belief that energy cannot be destroyed, it cannot be created, but that it can only change in form and move from one form to another.

Then one will have to ask, what are the qualities of energy that force it to remain indestructible? What is the ground zero form of energy that defines all other energy? The cymaticists are postulating that sound is the unifying and quantifying form of energy and that it is the only known force that can create something out of nothing.[93]

It is easy to conjecture that pondering this whole dilemma of matter and space and time and beginnings triggered the quest for the next level after the atom—once treasured as the smallest particle of matter—was split. Ernest Rutherford is recognized as the progenitor of nuclear physics; it was he who first suggested that there were smaller particles within the atom. Robert Oppenheimer carried the research farther and in 1945, the atomic bomb was exploded for the first time. Such extreme scientific exploration is, to say the least, daunting, and caused no small amount of trepidation for Oppenheimer. **He remarked—that it brought to mind words from the Bhagavad Gita: 'Now I am become Death, the destroyer of worlds.'**[94]

The driving urge for knowledge and resolution never flags. Science had discovered that there was another entire world of matter and wonder beyond the veil of the atom—protons, neutrons, electrons, gluons, and that "something" held them in perfect order and balance in their world, just as "something" held the hugeness of the planets and stars in their orbits as well.

---

93   https://www.facebook.com/3dcymatics/posts/559156050825418.

94   https://en.wikipedia.org/wiki/J._Robert_Oppenheimer.

Act 2 — Scene IV

In 1984, a group of scientists meeting in Paris, France, grappled with a growing new idea: String Theory. Among other postulates, it said:

> "If String Theory is correct, the entire universe is made of oscillating strings—Perhaps the most remarkable thing about String Theory is that such a simple idea has such a lot of merit. String Theory encompasses Quantum Mechanics, The Standard Model (which has been verified experimentally with incredible precision) and Einstein's General Relativity (gravity). The strength of String Theory is that it accounts for all four known forces in one elegant theory. But it should also be noted that to date there is no experimental evidence that string theory is the "basic description of nature."[95]

As a musician, I was astounded by the analogies that they, the string theorists, gave to describe their theory. The only workable comparison for what they were thinking was musical in nature!

> "Many physicists when explaining particle strings compare them to musical strings, as in a violin or other string instruments. In music, different sounds can be produced depending upon where a string is plucked. Different vibration patterns from a violin string each produce a unique sound. Similarly, in String Theory different vibrations from the infinitely small 'rubber bands' result in particles with different characteristics. The strings vibrate at

---

95   http://www.particlecentral.com/strings_page.html.

resonant frequencies. Every type of string has a unique resonance or 'harmonic.' Some harmonics determine a fundamental particle (as opposed to a force) whose mass and other characteristics are defined by the string's resonant pattern. This is in stark contrast to the Standard Model whose many particles have characteristics derived mainly from experimental data. String Theory encourages us to think of the vibrating strings not only as dictating the properties of the particles, but as 'being' the particles. The same string harmonics idea applies to the four forces of nature as well as the particles. Different harmonics determine different particles and forces so that all matter and forces are united in one elementary theory. The characteristic patterns of vibrating, oscillating strings provide the music of science."[96]

Another random thought: String Theory cannot be proven at this point in time. Why? Because it delves into worlds that are so infinitely infinitesimal that they cannot be seen, tracked or observed in any way. **"Therefore, if we are ever to prove that strings really do exist they will have to be 'inferred' from other data."**[97]

The theorists know this, of course, and are undaunted by the quest. Random thoughts have staying power: they feed one another. It is a wise theorist who understands this and will allow his or her research to benefit from the research of other "disjunct" disciplines.

96 http://www.particlecentral.com/strings_page.html.
97 Ibid.

And here is precisely where the research of the emerging science of "cymatics" enters the picture. The mystery of vibrating energy has always intrigued science. Greek philosophers were already toying with the idea that sound was the result of vibration in 500 BC. By 1500 A. D., da Vinci was making the assumption that sound did, indeed, travel in waves. Galileo continued the quest and for the last half-millennium, knowledge has grown exponentially in this arena, until we have finally arrived at this point: could it be that vibrating energy, the essence of sound, the essence of music is, in fact, the cohesive force that holds the universe together?

John Stuart Reid and his colleagues in the UK are in exciting pursuit of proving the efficacy of the statement that "Everything is sound."

This brief excursion in random thought is meant to simply paint a word picture, which will be ongoing throughout the remainder of this book, with the secondary agenda of illustrating the connectedness of all things—and the reason for that connectedness from a logical and **peripherally** theological point of view.

SOUND TAKES RANDOM PARTICLES AND SHAPES THEM INTO FORMS. SOUND IS ENERGY WE HEAR MADE BY THINGS THAT VIBRATE.

The wonder of evolving knowledge is this: a random seed-thought is dropped into the mind of man. He acts upon that thought and one of two things happens. He makes an appropriate response, or—the thought itself triggers another random thought. If he makes an appropriate response, new discovery probably ensues. If another random thought is generated, the resulting new discovery may be delayed but in all likelihood the resulting new discovery will be even greater than expected.

119

And, perhaps, it was so with Ernst Chladni, as he responded to the queries set forth by Aristotle, Fibonacci, Galileo, or da Vinci hundreds of years before his time. Curiosity is triggered by random thoughts, and it may take hundreds of years and many generations to finally come to a logical and acceptable conclusion. Thus, it is that the most astounding discoveries come on the heels of such thought and query, and, indeed, it is an evolving landscape of idea upon idea, imagination upon imagination, possibility upon possibility that finally takes shape in a sound and acceptable theory. And this may be analogical, too.

So, Chladni scattered salt or sand on a copper plate of specified dimensions, bowed the plate with his violin bow and made vibratory sound which, in turn, caused the particles to dance into very specific shapes on the plate. And it has taken these past three hundred-plus years for the idea of vibrating sound as the creative/cohesive force in our universe to come into its own. Only since the late 1990's has the newly emerging science of cymatics become a recognized discipline in academia. Even so, it is not yet fully acclaimed.

The building ground swell of evidence in this new and emerging science will, I believe, soon compel an audience which will give it full acceptance and will begin to respect its findings and use them to the advancement of the general knowledge of life, creation and our universe(s). There is to date so much adverse teaching in these fields that is high time for one undeniable set of facts and findings to finally make an impression which will turn our entire understanding of matter, time and all of creation into the one simple, beautiful unified theory that Einstein so doggedly pursued.

The concept of sound, words, song and frequency and vibration is central to most of the thinking about mankind and his cosmos today.

MATTER IS GEOMETRY IN MOTION

I can recall that my years in high school math, especially geometry, were most confusing and totally irrelevant to my understanding of learning and schooling. They were obviously non-essential and had absolutely no useful meaning to the student who was just as obviously a creative genius (as I figured, "I must be for I did not fit any other place"). Just as long as I could count the meters in my music, and as long as I could use numbers for the purpose of buying and selling, what more could I need?

Over the years, I would come into contact with curious things that never failed to pique my curiosity and set my inevitable dialogue with my imaginary friend, Jana, into fast-forward. For instance, an article in the *New York Times* about music and math as two potentially life-changing forces extant in the cosmos caught my attention back in the 1970's. Edward Rothstein, writing for the *New York Times*, cited mathematicians of by-gone days who were convinced of the necessity for balance and beauty in the field of math:

> "G. H. Hardy, added to the building awareness of—relatedness when he wrote of the harmonious character of mathematics: "Beauty is the first test; there is no permanent place in the world for ugly mathematics." He no doubt inspired another mathematician, Henri Poincare,' to write about "the feeling of mathematical beauty, of the harmony of numbers, of forms, of geometric elegance."[98]

Rothstein, in the same article, rekindled my interest in the philosophy and writings of Pythagoras. His philosophies embraced mathematics, music, medicine and even theology. *Encyclopedia Britannica* comes to the point in one statement about this ancient Greek scholar:

98   *New York Times; Math and Music: The Deeper Links*: Edward Rothstein: August 29, 1982.

"According to tradition, Pythagoras (c. 580–500 BC) worked in southern Italy amid devoted followers. His philosophy enshrined number as the unifying concept necessary for understanding everything from planetary motion to musical harmony."[99]

The ramifications of Pythagoras' intellectual excursions are too numerous to deal with in this writing. Suffice it to say, however, that even a cursory investigation would soon reveal that his influence in the world of knowledge of all kinds is incredibly vast. Out of his theorem, $a2 + b2 = c2$, came the Golden Triangle (and the Golden Mean), the western musical harmonic system, and all variations of architectural and artistic structural configurations.

The wonders of geometric logic had just begun as Pythagoras and his followers followed their stars—and that process is still in process! And will probably remain in process as long as there is time.

In 1981, Gyorgi Doczi published an incredible book called *The Power of Limits—Proportional Harmonies in Nature, Art and Architecture*.[100] In the preface he makes it clear that he is seeking the same answers to the same questions which have shadowed scholars for millennia:

"This book searches for some of the basic pattern-forming processes that, operating within strict limits, create limitless varieties of shapes and harmonies. It is an inter-

99 http://www.britannica.com/topic/number-theory.

100 *The Power of Limits: Proportional Harmonies in Nature, Art and Architecture*: Gyorgy Doczi: Shambala Publications, Inc. Boulder CO: 1981.

disciplinary venture into the no-man's-land between the borders of science, art, philosophy and religion, an area that has been largely disregarded in recent years because its contents are intangible. This area, however, bears investigation, since the powers that shape our lives and our values have their **source** here."

Doczi had his fingers and his mind on the pulse of all nature and saw, therefore, the acute relevancy in the idea of connectedness that swirls about the intelligentsia of every age and discipline. He came very close to recognizing **and naming** the Power behind the beauty of the unity that he so eloquently described. It is one thing to "discover and re-discover patterns of order and beauty in nature"—it is quite another thing to make that leap of faith, that leap which cannot be proven to be the next truth, and thus understand that there is an Almighty and Divine Designer Who is responsible. And these designs are incredibly undeniable in themselves: they are revealed by "slicing through a head of cabbage, or an orange, the forms of shells and butterfly wings. These images are awesome not just for their beauty alone, but because they suggest an order underlying their growth, a **harmony** existing in nature. What does it mean that such an order exists; how far does it extend?"[101] Indeed. How far can we legitimately go in ratifying the idea of Almighty God, the Three-in-One, the Eternal Father-of-All, and Divine Designer, Whose fingerprint is on all that is made?

Doczi's thinking is paramount to the theory of creation as a musical and Divine event, because he portrays such a clear view of the unity and

---

101 Credits written by publisher: "The Power of Limits."

connectedness in all of the universe. He seems to understand the laws and parameters that govern structure and matter, and he is successful in making that understanding accessible to the common man. What is interesting to me is that most, if not all, of these thinkers and writers fall just shy of seeing that **precisely because** there is this degree of connectedness in all of nature and the cosmos, there must be, by definition, **One Designer.**

Sidebar: Interesting—for the last three years, ever since I have been a subscriber to online satellite radio, classical music has been my constant companion at the computer as I write. Not that kind of deliberate, directly involved listening that I do at other times, but simply the kind of companionate listening that gives a background of order and beauty (vis a vis my own theory of music as the raw material of creation), that is an ongoing aural composite of the same kinds of things that I am writing about. In the process, I often hear a composition that is new to me. However, there is an immediate response to that new sound: "Oh, that must be an English composer—probably Elgar." Or; "Hmm! That could only be Shostakovich—early Shostakovich, at that!"

What does my brain know instinctively that I have not realized as yet? It "hears" the hallmarks of that designer, and it can spontaneously decipher the colors and leitmotifs that identify that designer—the composer and his/her fingerprint. And furthermore, it can "hear" the nationality colors and styles, as well. It is a very exciting and powerful insight.

EVERYTHING IS SOUND. IN THE BEGINNING WAS THE WORD—AND THE WORD IS SOUND. FROM FORMLESSNESS COMES FORM—JUST ADD SOUND.
John Stuart Reid

Act 2 — Scene IV

Kepler said: "At ubi materia, ibi Geometria"—which means, "Where there is matter, there is Geometry." He went on to say: "I am merely thinking God's thoughts after Him."

Kepler was a devout Lutheran, and so it is appropriate for him to think in terms of the Almighty. Perhaps we of the 21st century have lost a great deal of that devotion to truth in homage to the enlightenment and ascendancy of mankind.

To create little boxes in which to store the elements of the truths that are uncovered along the way on such a journey as writing this book and the one which preceded it, *God's Song*, is an impossibility. One simply spills its contents over into another. So here is another of those points at which one could and probably should engage one's alter ego—one's "Jana" to help sort out those contents. It is a theological consideration to differentiate between the words, *"In the beginning was the Word and the Word was **sound**,"* and the true origin and transliteration of that phrase—*"In the beginning the Word was with God and the Word was God."*[102] Remember, it is neither my intention nor my desire to practice theological apologetics in this book: it is my intent and my desire only to present my beliefs, which may then be taken up by more qualified thinkers and writers than I and molded into a theory that is worthy of being accepted as a fully developed theory. I shall defer that consideration until that ACT dedicated to FAITH AND THEOLOGY.

That in no way detracts from the importance of and the anticipation with which I view the findings of the emerging new science of cymatics. I was first introduced to cymatics about a year ago. According to my initial research, John Stuart Reid and his colleagues became really active

---

102   John 1:1-3 New King James Version: New Testament.

in their work only a couple of decades ago. The catalyst which brought together some of the most dedicated scientists from several fields of endeavor was the near-death experience of Lynnclaire Dennis in the UK. She was responsible for bringing together a team of forward-moving and forward-thinking scholars from the UK and the US, including John Stuart Reid from Cambridge, UK, Jytte Brender McNair of Aalborg University in Denmark, and Louis Kauffman, University of Illinois Departments of Mathematics, Statistics and Computer Science.[103] This book was not conceived, as I understand it, to promote a thesis concerning creation, but, nevertheless, their research is exciting, valid and profound, and I believe it will be instrumental in revolutionizing our concept of our universe in many, if not all, disciplines. I do not find the actual graphic and concept-driven ideas and images, which they are testing and portraying, to be directly opposed to our Christian concepts if closely examined and properly used. In fact, I find it interesting, that even though they make it very clear they are a secular endeavor, the things they have already uncovered are, in fact, quite in keeping with what I see as the truth about creation. Truth is its own defender, and if there is, as I believe, truth in their research, it will not go unnoticed or unrewarded.

What they have already found is that there is a plethora of related shapes and that these shapes are directly the result of vibration and each frequency has its own unique shape to reveal. String theory has known this for many decades now, but they have not yet found a way to empirically document what they know. I believe that the cymaticists will, given time and attention, be able to solve that need.

103 *The Merion Matrix*, 2013: Elsevier Inc., 32 Jamestown Road, London NW 1 7By, UK: Copyright c 2013 Elsevier Inc., All Rights Reserved.

Act 2 — Scene IV

The following quote from one of their websites is a must for your lists of things to investigate:

> "Cymatics reveals a strange and beautiful symmetry at work in nature. Inspired by the work of Dr. Hans Jenny we create images using water and sound. Our images mirror the symmetries found throughout the natural world, from the hidden shapes buried within snowflakes to the massive hexagonal cloud formations found on Saturn. Cymatics.org is a gallery of our work and a resource for other artists."[104]

By this time in the unfolding of man's existence and quest, it is becoming an imperative: to find truth as it is—in truth.

---

104   http://cymatics.org/.

Scene V

WHAT DOES COME NEXT?

SEGUE:
From the Italian: There follows, from the Latin segue:
First Known Use: Circa 1740

DEFINITION OF SEGUE:
1: to proceed without pause from one musical number or theme to another;
2: to make a transition without interruption from one activity, topic, or part to another.

*W*hether or not one takes Genesis, the fourth chapter, as undeniable fact or as beautifully written myth or allegory, the impact of the story remains undaunted. My own personal insights dictate that I take the Scriptures as undeniable. But individual interpretation does not change the verity of the creation as being a musically understood feat of such proportions that only a Divine Designer, Almighty God, could

have designed, orchestrated and then, realized in perfect detail. Before the first astonishing exchange between the Creator and the created took place, the only sounds familiar to the ones who inhabited the earth were the sounds of their own voices and, possibly or maybe even probably, the voice of the Creator.

And remember, the physical, emotional and mechanical aspects of our earthly lives have been around now for thousands of years, and the natural outcome of such ongoing familiarity is to take those things for granted as if they had always been. Not so. Scripture can be very exciting and intriguing, if read and pondered imaginatively and with care. Genesis 1:1-2 reads: *"In the beginning God created the heavens and the earth. The earth was formless and void, darkness was over the surface of the deep—and the* **Spirit of God was moving over the surface of the water.** *"*

Imagine! It has long been accepted that Genesis is a book of beginnings, perhaps the greatest of which is the beginning of expressed truth, available for man to understand. The vastness of creation can only be understood in increments—the full panoply of grandeur is far beyond our human finiteness to take in all at once. But just consider the enormity of the earth and the universe in a state of complete void and darkness. I can also, in my world of Jana and making the incredulous more credible, feel the **silence** so deep and so thick that it is more felt than audibly perceived—and **then**, *"God said."* "God said" or "God sang." The **very first sound** among the multiplied millions and trillions of sounds that have been uttered since? Was God's voice, in truth, the very first sound?

You see, from here on in time, this entire universe of things recently created is now in need of observing itself and learning—being taught. It is in need of learning all and everything that we now know and take so for granted, as said before. The main idea here, (and it is, indeed, a transient "half-idea" of the sort with which Robert Shaw tantalized his people in the study of the great choral works),[105] is that whatever one may be conscripted to believe about the timing and mechanics of the creation event, truth remains truth, and the truth of the Biblical account is inviolable. Interpretation of truth is a moot, and maybe even unsolvable, question, but the truth of the **event** is not what is being questioned—at least not in this writing.

In that period of void and darkness, there was a totally empty, blank score upon which the Creator would record His inexplicably vast and beautiful work of musical art with all of its ramifications, colors, harmonies and deeply related systems and mechanics. To comprehend the extensiveness of that creation is a mind-crushing exercise—virtually impossible, but to apprehend the event itself and realize that it was accomplished for His creatures' benefit is a good point with which to start.

If one can take oneself back in time to that point when there was nothing but void, whatever that may mean, and allow oneself, then, to see the present in terms of what there is now, the task may be made somewhat simpler. For instance, when there was nothing, the first thing that happened was a sound. Was that sound the voice of God? I believe that it was. And given that my entire life has been spent as a musician, and given that my faith is at the very source of my thoughts and theories about creation, hearing that voice as the almighty voice of the Almighty God is a simple matter. And that is precisely why I urge the reader to consider these "half-ideas" as ventures of sanctified imagination so that they do not become cluttered with the dust and contamination of our present world. Another "half-idea" that has grown into feasible reality for me is the idea that the voice of God could be heard as being spoken—or it could be heard as being sung. The Hebrew language supports that possibility.

---

105 *Dear People—Robert Shaw*: Joseph A. Musselman: Indiana University Press: 1979. pg. 72.

The first theme was the theme of creation itself, in the midst of void, when "God said," and now, thousands of years later, we are beginning to see the emergence of empirical research which says, "Vibration underpins all matter in the universe. No matter can exist without sound and vibration."[106]

Now, if we "fast-forward" to the present and to the fascinating findings of science couched in the disciplines of string theory and cymatics, we have a distilled idea which is no longer a "half-idea." This is the second theme, and the presently critical one. To see that what was put into motion "in the beginning" is now coming to fruition and is even in the process of being empirically documented. It is a huge leap of time and knowledge, and of contemplative debate.

Imagine the sense of wonder that Jubal must have experienced when his Creator told him that he was to make a flute from a reed-pipe and learn to play on it. And, as if that were not enough, the Creator God then told him to build a harp, tune it and learn to play that as well.[107] And that was only the very beginning of eons of time and expansion, intellectually and spiritually, among people of growing number and accomplishment—almost too much to contemplate at all, much less to be able to comprehend the ramifications and implications of that immense development.

But now, at this point, we need to move on into the world of today, where science is commencing to define new ideas about so many things such as sound and creation, vibration and matter, and is beginning to see connectedness where connectedness was ignored before and where it went unnoticed and neglected for most of time-extant. Now is an age of excitement in many fields of research where new ideas and "half-ideas"

---

106 *God's Opera*, ACT I, Scene I.
107 Genesis 4: 17-22: New Testament NKJV.

are surfacing and the people who are closest to the center of those ideas, like the string theorists and cymaticists, are embarking on new lines of thought and experimentation. They are capable of more distant excursions into new knowledge because, in this age of technological augmentation, they have much more at their disposal to aid them.

However, not everyone is in a place conducive to accepting the theological along with the scientific, but I do not see that as problematic. It is a natural result of secular thinking and there is, so far, little apparent questioning of the relationship of such vastly differentiated disciplines. At least they are differentiated on the surface. I believe all that is needed is more time to ponder—and perhaps more scholars to catch the vision of the possibilities which will present themselves as a result of considering that the entire universal creation cries "unity"! That is the actual meaning of the name "universe." Yes? And—that the raw materials of that universal creation can be explained spiritually and musically. Let's explore.

1

***Proceed without pause from one musical number or
theme to another.***

2

***Make a transition without interruption from one activ-
ity, topic, or part to another.***

It is hoped that the reader will continually be brought back to the idea that was irrevocably lodged in the mind of Einstein—that there is one simple, beautiful set of laws that will explain creation and the control of

the entire universe. He came very close to the truth in his quest—maybe, in the depths of his own soul, he did see the truth.

Einstein said,

> "I want to know how God created this world. I am not interested in this or that phenomenon, in the spectrum of this or that element. I want to know His thoughts; the rest are details...Science without religion is lame. Religion without science is blind."[108]

In the process of researching to find as much of the truth about our universe and its contents as was possible for me to find, I have been drawn back again and again to the overflowing wisdom spilling from the mind of Albert Einstein. His seems to be a humble approach to all things, and that is quite astounding for one so brilliant as he. He did not confine his efforts to the laboratory or the study desk. In fact, I suspect that he probably believed that the entire created world was his laboratory and the very ground on which he walked was the desk of discovery.

One could, I suspect, acquire an entire education by making a study of his slightly casual quotes. I say "casual" because he seemed to be able to saturate a 25-word sentence with volumes of meaning. And what seemed like random ruminations were pointed discoveries voiced eloquently and yet understandably. And his convoluted thinking was not limited to mathematical or scientific pursuits. He could be quite philosophical. For instance:

---

108 http://www.simpletoremember.com/articles/a/einstein/.

"The scientists' religious feeling takes the form of a rapturous amazement at the harmony of natural law, which reveals an intelligence of such superiority that, compared with it, all the systematic thinking and acting of human beings is an utterly insignificant reflection.

"There is no logical way to the discovery of elemental laws. There is only the way of intuition, which is helped by a feeling for the order lying behind the appearance.

"The intuitive mind is a sacred gift and the rational mind is a faithful servant. We have created a society that honors the servant and has forgotten the gift."[109]

If one were to make such a study, I believe that one paramount thing would surface almost immediately: the child-like functioning of the immensely intellectual mind has at its core the **intuitive—the imaginative.** Einstein must have believed this as well, as marked by another of his "casual" quotes:

"The most beautiful thing we can experience is the mysterious; It is the source of all true **art and science.**"

This is a particularly revealing statement. Without making the overt point, Einstein makes it clear that art and science are somehow related. A profound thought to ponder.

What does this all have to do with the consideration of **sound and the science behind that sound**?

109 http://www.simpletoremember.com/articles/a/einstein/.

Some of the most revolutionary research and discoveries of all time have been made in our lifetime. The age of technology has exponentially expanded our fields of knowledge, and, not only that, but that same technology has put knowledge at our fingertips, which in another era would not have been accessible except to the very few, the intellectually elite who were fortunate enough to be able to afford something like a Cambridge or Oxford education. And even then, research became a burdensome task, what with huge volumes written eons ago and then hidden away on cavernous library shelves deep in a cold stone edifice. In this remarkable age in which we find ourselves, anyone with a computer and a sense of curiosity can learn about anything that finds its way into his or her field of vision.

And so it has been with me as I sought to unravel some of the mysteries of creation. I have had a somewhat different approach, I think, to making the relevant points that commanded my attention. In no way have I felt it to be my task to change the course of any of the rivers of knowledge that are flowing so fast in our age. Nor have I had in mind to challenge or to refute any of the thinking of the scholars or bards of this present time or the past. My quest has been quite simple: the quest for truth—which started as a musical quest but which led in so many surprising and gratifying directions. True, I was dedicated to pursuing this quest from the perspective of the evangelical, Scripture-centered point of view, but not in the sense that I wished to "convert" or in any other way openly influence my colleagues who are also in pursuit of truth.

Truth has its own strength, and will in its own time, make that strength known and thereby successfully reveal itself—"truth." Of course, being a musician and a life-long observer and student of music, I have a vision that may not have occurred to many—that there is truth—real truth—in music. A recurring leitmotif is just that: the presence of truth in music.

Act 2 — Scene V

In the beginning, it was merely a half-idea, transient in persona and character, and, as Robert Shaw so often lectured, a necessary thing to understand. Yet, when we would ask him what he meant by "truth in music," he would shuffle his feet, chuckle and say, "When you find out, let me know." But, he was convinced that it existed—"truth in music." Perhaps, this is the way it must go—

It is probably evident by now—at least to some of the readers—that my reason for writing this book is two-fold. First is to depict in some kind of an organized fashion the theory that I hold, and hope to propagate, that God "sang the creation commands." Second, but perhaps more important, is the desire to open a new window into the profound beauty of that creation and its manifold systems, creatures and natural wonders, and to bring into sharp focus its enormous consistency and connectedness.

Then, I had to consider what His raw materials were, because Scripture is quite unequivocal in its claim that creation—all the worlds, universes and systems imaginable—were created from **nothing**. For eons of time, that stance was problematic. But, in recent decades, there has been a growing scientific awareness of the presence and validity of the role played by sound—vibration, frequency, periodic function—relative to matter and the constancy of the galaxies and all that relates to them. And, of course, sound is not in any way tangible—or is it? At least, we can safely say that sound is not material.

This is an interesting and compelling departure from the norms we have held for so many eons. Now, science is literally making a way to see, photograph and study sound as a viable entity—the world of cymatics—but it remains a non-material entity.[110]

---

110 Cymascope.com: John Stuart Reid: All rights reserved: 2008-2015.

An imaginative thought: the creative power and energy of the Creative God suddenly utters a magnificent, harmonious and all-encompassing sound with the intent of creating what He is imagining—worlds, stars, planets, vegetation, mankind. This is precisely what I believe Tolkien and Lewis were imagining when they put pen to paper and wrote of Iluvatar and Aslan, the singing creator gods of Ea and Narnia. The "great music" which the creator-god gave to his thought-music-beings was not given all at once. He gave it theme by theme, harmony by harmony, leitmotif by leitmotif until the sound was overwhelming and all-encompassing. And then it was finished!

How like our earthly experience under the caring creative power of Almighty God. He has not given us that all-encompassing understanding of the wonder and mystery of our existence and the reality of our planet and universe. Little by little He exposes His mind and His creative tools. I am astounded by the immense complexity of our universal home and the expansive breadth and depth of the galaxies to which we belong. But even more astounding is the confounding simplicity **and unity** in which our complexities are couched. And the more knowledge that is revealed to us, the more "simple and beautiful" the governing laws become to our faltering minds. This is the "truth" that Einstein was in quest of. And since the possibility of a "Singing Creator God" is becoming more and more feasible, the more accessible the possibility of "truth in music" becomes. How logical it would be for the Creator God to reveal Himself in that mode—music.

But there are several caveats that keep recurring in their insistence to be heard—one of the most insistent is that we are not talking about music as we—the created—envision it. Even Ilvatar cautioned his Ainur against thinking that they could compose such magnificent creative sounds—

those were his domain. And so it is with us, as well. Our creative genius is only as God Almighty sees fit to grant it, and the result of that creative genius is but a smoky, dim, mirror image of the greatness of the Creator.

There is a captivating takeaway resulting from this line of thinking: the worlds of art and science are neither confrontational nor contradictory, and, by inference, the world of theology is slowly taking on a more compatible aura of truth and reality in the scientific sense. That is, if one is deeming theology as what it really is—the study of God— and, in particular, the Creative God.

Recitative

In the opera, a RECITATIVE is defined as the 'recitation.' A character in the story tells what is going on in either a spoken narrative or an arrhythmic line, accompanied by a keyboard instrument. By that definition, it is easy to see how it departs from the basic structure of the musical format, the harmonies are less strict because of the more conversational style of the narrative, and it serves it's purpose of furthering the story line without interfering with the general scope of the operatic story. In fact, the recitative acts as a cohesive pattern in that it can elaborate upon the story line, clarify interactions and explain emotions, provide information that may be otherwise too vague or even too complex to include in the main body of the opera.

Jana has been this kind of imaginary entity for me over the years. She can go to places in the privacy of her great imagination that then allow me to take that flight of thought and vision and shape it into a viable hypothesis to present in a more acceptable fashion. Remember: "Imagination is more

141

important than knowledge," according to the great Albert Einstein. But lofty imaginations can be difficult to accept or manage for mankind at large.

Following is the recitative which I offer to give insight into the writing of this operatic analogy which I designed to, perhaps, give shape to the massive justification for and the tools used in this creation chronicle.

So far, in this operatic creation narrative, this is what we have. We have a simple question that was first posed many, many years ago: What is music and from where did it really come? That apparently uncomplicated question soon spawned a much broader and much more demanding question: What is the real truth about creation and the origins of all life? Is there any connection between the two quests? I concluded that there **was** and **is** a very deep connection between the two quests—the origins of music and creation itself.

However, Jana and I were quite satisfied for many years to simply entertain the origins of music. There was certainly enough there to keep us thinking and studying and researching and contemplating. Music by itself provided a rich field in which to dig and plant, sow and harvest, such as the ages of musical development, the nuances of rhythm as it proceeded in the growth process, the richness of expanding harmonies and the burgeoning of thousands of styles and structures—modal harmonies, major harmonies, minor harmonies, Western harmonies, foreign harmonies. There are also the nationalistic musical dialects (musical dialects common to the passing ages of history), types of instruments of such breadth and variety that it almost becomes mind-blowing, composers who gave themselves to developing beauty, and composers who deliberately gave themselves to trying to dissolve what their predecessors had built.

Underlying the crazy quilt of this magnificent display, the same question kept manifesting itself, in more and more detail, in greater and greater depth and breadth: But from whence did it come, and to whence was it going? Why this glorious obsession in the human landscape—the obsession of music and its unfathomable diversities and similarities? What is its actual intent in the complete scheme of things? The quandary seems almost impossible to even think about, much less to solve. Is there really any reason to be found for its myriad shapes and forms and uses?

It wasn't long after I started ruminating over this fairly well-focused quest concerning music and its origins that I began to realize that the real power of music lay in finally comprehending that music in and of itself was only the tip of an iceberg so gigantic that it was terribly intimidating to consider. Music was so omnipresent that it could not be entirely expunged from any other discipline. If one were to be honest with one's self, one would have to allow oneself to grasp the presence of music in everything: astronomers could hear the music of the stars; doctors could comprehend the presence of music in the human mind and body; mathematicians could easily grasp the musical balance and order in their equations; physicists most often use musical analogies to explain the deepest difficulties of their theoretical expeditions; ornithologists may know best the beauty of music in nature, as they identify genus after genus by the song sung; certainly, the greatest poets and playwrights were forced to deal with meter and contour in their writings. Light waves can translate into sound waves; Pythagoras discovered the astounding relationship between frequency and vibration and sound and mathematics. I dare say, one could not find a single area of knowledge where there is not, at least, an inference of the presence of things musical. Consider the Aurora Borealis.

There had to be a huge and viable reason for the omnipresence of this force called music. And, just perhaps, the reader may begin to grasp the idea that it could, in fact, be that this thing called music, in its most pure

and eternally conceptualized form—not the form comprehended by the finite mind of man—is the actual stuff of creation itself—the raw materials, if you please.

It was the discovery of string theory that jarred me into the pursuit of truth in a new way. I had long since entertained the idea, without the aid of my sacred imagination and alter-ego thought-friend, Jana, that there was vastly more to the study and understanding of the full impact of music and its power than any writers up to the 20th century had apprehended.

There, also, was a concept niggling around in my head that I could not quite get my mind around: TRUTH IN MUSIC. Robert Shaw had always taunted us about the idea that truth was inherent in "good" music. He never defined "good" music, and when asked what constituted that "truth" in music, he always replied, "When you find out, please tell me." But we knew that he intensely believed in the presence of "truth in music" and felt that it was elemental.

When I heard Brian Greene, of Columbia University, describing string theory, the elation I felt was palpable. Of course! The basic tenet of string theory, in lay terms, is that ALL MATTER is finally comprised of infinitesimally small "strings" of vibrating energy! Vibrating energy—the express essence of music!

That sent me on a two-year exploration of physics and subatomic elements. Being no scientist, I had difficulty understanding enough of this heavily weighted theoretical science to be even marginally intelligent about it—but it worked!!!! It fit!!!!! Amazing!

And the more I learned about quantum physics and relativity, Einstein and Kaku, and the newly emerging science of cymatics, the more convinced I was that there was something very evasive that no one had up to this point, investigated—and the cymaticists[111] were now putting into practice and into words their own understandable and empirically demonstrable experiments that gave them, at last, the courage or insight or both, to say that without sound, nothing could exist. Sound is the cohesive force of nature. The haunting idea that all things are related and that there is cohesiveness and connectedness in all the universe was actually beginning to take shape.

In the forty-plus years that I have been thinking about the universe, the raw materials of creation, the almost limitless scope of the presence and influence of music in that universe and the myriad of inferences of such thinking, I have yet to find another author who gives credence and detailed study to the idea that all things are, indeed, connected in their genesis and that all things are, indeed, related—one discipline to another, one structure to another, one concept to another. That is the genesis of this writing—to attempt to draw the parallels; to attempt to delineate the similarities, to draw attention to the immeasurable beauties of a creation that carries the thumbprint of an Almighty Creator God Whose template is as consistent and perfect as He Himself is.

The providential power of language was basically discovered by chance in my quest. I started to study Biblical Hebrew as a means for doing my own linguistic research in the Old Testament. It quickly became obvious that this language was far more than spoken communi-

---

111　My footnote: The work of the scientists who are on the cutting edge of the emerging science of "cymatics" is discussed in ACT II, in the second and fourth scenes. It is my personal conviction that in time we will be able to observe the results of years of experimentation in the field of "string theory" made visible by the work of this group of learned and courageous people who are front-running making sound and its component parts visible. Furthermore, they are illustrating that the forms and shapes that we have been familiar with for decades in geometry, knot theory and architecture are the same as the shapes of sound made visible. Amazing!!

cation; it was a living thing. How could this be? The professors will tell you that Biblical Hebrew is a changeable language and that its myriad meanings have to do with the context in which they are found. That is true, but it is far more than that. Little did I know when I enrolled in this class at Liberty University that this was to become not only a learning linguistic experience but also would be an intellectual adventure which would teach me far more than a new language and would eventually revolutionize my quest.

The first surprise came when I started learning vocabulary. One of the first verbs I learned in my naivte` was the word *dabar*. I call it naivte,' because I had no idea at that time that Hebrew contained the thousands of nuances and hidden marvels that it did and that this word was simply one isolated example.

The marvel of that word, *dabar*, was that it meant both "to speak" and "to sing," and included in that repertoire of meanings were "to create" and "to command." But to have the dual responsibility of "speaking" and "singing" was miraculous to me, for I was beginning to realize that "truth in music" had a more eternal facet than I ever could have conjured.

I did not know about rabbinic traditions which delve into myriads of secrets: secrets of the numbers which are relegated to each alphabet letter and which have incredible meaning all of their own: secrets of the graphic integrity of each letter—of each consonant. I did not know that each Hebrew character is a graphic depiction with meaning—a picture, if you will, of the particular thing to which it may refer.

It was not until I met Rabbi Mordechai Kraft online that I suddenly realized, much to my own chagrin, that it was obvious that God had to teach Adam and Eve (yes, I believe in the literal account of the Garden of Eden) how to communicate verbally: that necessitated a language. Just as obviously, it was not difficult to understand that God, therefore, was the Designer and Writer of that language. Hence, that language is most detailed, perfectly designed and immensely intuitive and complete.

If one would give oneself time to explore some of the wonders of Hebrew and how perfectly its numeric and linguistic patterns fit and enhance and support each other, I believe it would be life-changing—as it was for me. And, the incredible sidebar is that there is an incredibly musical component in the language itself. No accident. No coincidental advantage. Remember that people of the earliest culture, could not read and could not write.

All knowledge was conveyed verbally and audibally. It is now an established psychological fact that words which are sung have a more lasting staying-power than prose simply spoken. Hence, the advent of the point system of the Hebrew language which represents a pitch change as well. By and by, the tabernacle priests became literate and began singing the Psalms and other Scriptures in their worship. As a matter of fact, the Book of Leviticus clearly documents the appointing of the singing priests by Yahweh as the worship structure of the Hebrew people grew.

And so the chronicle progresses and unfolds. A simple intellectual inquiry led eventually to revealing the connections between music, all other disciplines, if you please, and especially the discipline of science. And then of language.

Next, we must address one of the most salient and probably, most difficult, of disciplines—theology within the framework of spiritual power and influence. Here, there will undoubtedly be issues that will conflict with many of the mores and traditions of the readers at hand. It never has been the intent of this writing to defend or provide apologetics for theological issues or scientific issues either.

147

However, I must make it clear one more time that the basis for my entire thesis is Scriptural, and I think I must admit to being a literalist in that realm. I am not an evolutionist, by any stretch of the imagination, and I am not ashamed to admit to being a Creationist. I cannot but hope that the utter and complete diversity and yet the utter and complete order of the universe and entire creation will bring the reader to ask, "How and why? Can it be accidental—coincidental?" The tremendous amount of lavish creativity which comes out of such a few basic laws and guidelines must at least pique the imagination to say, "How can this be—without the presence of deliberate and intelligent design?"

Part of the consideration of this theological Act must be consideration of frequency and sound and vibration as it is connected to the divine personage of the Work of God, Jesus Christ.

And then to look finally at the wonder of music as an integral, intelligent gift out of which the Creator made all that is. Sound is the cohesive factor. Music is sound. Music has all of the elements from which all things benefit. Therefore, its ubiquity and all-pervasiveness can be better apprehended.

Along the way on this quest, it has been my great honor to see, to hear and to learn. To see, hear and learn what? To really see a snowflake, for instance. To really listen and actually hear the shifts in pitch caused by a single loss of vibration per second on a violin string. To learn why that principle worked and why it was the same principle that worked in both what I saw and what I heard. To really see deep into outer, outer space via the Hubble telescope and to wonder over the depth of perception that was provided by those Hubble videos. To turn the volume up and listen to the light waves being emitted from the planets and stars on the video and hearing them translated into unearthly music. To learn that there is really no difference in the principle of light waves giving me sight and color and sound waves giving me beauty that can only be perceived by the ear and the brain.

And then! And then, to realize that my brain, my ears and my entire body actually operate on the same foundations of vibration, frequency and sound that are operative in every single realm of creation and on every single level of activity. This is one of the issues that has confounded physicists for eons of time—how these equations and concepts and systems can all be related and really act as one dynamic and massive unit—called creation. At first, the entire experience was confusing and complex to the degree that it was almost impossible to assimilate. It has become far less elusive over time.

Act 3

THEOLOGY, FAITH & SCIENCE: A CREATIVE PATH TO TRUTH?

Scene I

A NEW TRIVIUM?

Theology *(is the) philosophically oriented discipline of religious speculation and apologetics that is traditionally restricted, because of its origins and format, to Christianity—The themes of theology include God, humanity, the world, salvation and eschatology (the study of last times). The concept of theology that is applicable as a science in all religions and that is therefore neutral is difficult to distill and determine.*

Science *(is) any system of knowledge that is concerned with the physical world and its phenomena and that entails unbiased observations and systematic experimentation. In general, a science involves a pursuit of knowledge covering general truths or the operation of fundamental laws.*[112]

Art *(is) the study of the nature of art, including such concepts as interpretation, representation and expression and form. It is closely related to aesthetics, the philosophical study of beauty and taste.*

---

112   http://www.britannica.com/editor/the-editors-of-encyclopaedia-britannica/4419.

he "new trivium," which I have intuited for the purposes of this book, is by no means any type of rebuttal against the aged and "tried and true" trivium of past scholarly endeavors. Of course not. It is simply a means of framing a logic for my theory and the consideration of the theology behind that theory.

When universities began to develop in medieval times, the subject matter was based on the concept of the "trivium"[113] or the "quadrivium."[114] The "trivium" was a group of studies consisting of grammar, rhetoric and logic which formed the lower division of the seven liberal arts. The "quadrivium" was that group of studies consisting of arithmetic, music, geometry, and astronomy which formed the upper division of those seven liberal arts.

The division of these studies into these specific groups is also of great interest. I repeatedly refer back to a scholar by the name of James Burke, who wrote the remarkable book, *The Day the Universe Changed*, in which he makes the profound observation that "…we are what we know and as the body of knowledge changes, we also change."[115] There has been a constant and prolific growth in knowledge down through the ages—especially since the "age of the enlightenment." One of the essential tenets of my entire theory of creation, music and the Creator God, is that nothing is accidental, incidental, or co-incidental. All things are connected and there are increasingly strong indicators that this is part and parcel of the truth which is being sought. The growth of knowledge has been according to a well-thought-out and perfect plan that supersedes our knowing. It is part of our basic quest for more and more knowledge and understanding to

---

113   http://www.britannica.com/topic/trivium.

114   http://www.britannica.com/topic/quadrivium.

115   Ibid. Pg. 62.

seek the real truth about ourselves and our universe(s). It is my hope that realization of this truth will become increasingly evident as this book progresses. Remember that one of the foundational stones in this construct is the idea of connectedness of all things.

Let's take a closer look at the "trivium" and the quadrivium." The trivium is stated as the lesser of the divisions of study, so the scholars of that day held. However, a deeper look at the identities of these divisions and their respective subject matter makes me wonder: grammar, rhetoric and logic. All having to do with "words" and the expression that is made possible only by **words**. Without these bodies of knowledge, the "quadrivium" would be hard-pressed to find expression.

Also, remember that science is now in the process of proving that vibration and frequency are the cohesive agents in all of creation, and the entirety of our universe is held together only by vibration and sound. According to the traditions and Holy Writings of ancient Biblical scholars, it was the "Word of God" that pre-empted all creation and actually was the "sound" that set in motion all of creation. The sad thing is that we human beings allow ourselves to be intellectually dwarfed by what we have been told over eons of time and have subjected ourselves, as well, to dwarfed imaginative skills that could, possibly, open our minds to exciting new trends of thought.

Just what if these early medieval teachers knew something that they could not quite express, but they saw the grave importance of knowing the skills and the art of speaking and rhetoric from the critical viewpoint that these skills were primary in all of creation? And what did they perceive in the study of logic? Maybe the preservation of a skill which would enable them to maintain critical thinking? And what has happened to critical thinking in our era?

The next tier of learning is: music, arithmetic, geometry, and astronomy. Why? Could these be the skills which point the way to understanding how creation really did occur and what those special studies which epitomized all the others could explain? What better disciplines than geometry, the study of shape and design and moving mathematics and **astronomy** which, in and of itself, necessitates the presence of and use of the mechanics of the other studies? It is my personal conviction that none of this is without great, infinite design and deliberate application of everything we know about basic knowledge of our world.[116]

Now, consider this: the beginning of learning in medieval times had at its heart "seven" disciplines of study. In the Hebrew language the number "seven" is considered the number of completion or perfection. Is it at all possible that the earliest scholars, who had far less information with which to cope, could see more clearly the importance of basic knowledge? It is, of course, inevitable that even knowledge would reach a critical mass as it grew and that the time would (and has) come when specialization becomes a stark necessity. But it is appearing more and more clear that a certain return to basics is also an essential. These seven bodies of knowledge contain all that is really necessary for us to grasp a beautiful truth: it is my belief that the content of these bodies of knowledge contain a secret, when wed to the Holy Scriptures, and when viewed through the exciting filters of modern technology that explains our origins and our universe's origins. In order to gain that perspective, it may be necessary, also, to be willing to lay aside our conviction that man's expertise lies at the center of all accom-

---

116 NB: https://www.youtube.com/embed/XRCIzZHpFtY This link was sent to me recently. One of the basic tenets of this book is the distinct connectedness of all things in our universal experience as human beings. If there were no master plan, no vision of completion, unity and order, it would not be possible to send a vehicle to outer space, to another planet and have returned to us in understandable form, pictures, documents and information relevant to our existence and the furthering of our knowledge of things important to us.

plishment and all advancement. The need for recognizing Almighty God does not deplete itself just because of our human egoism.

The "trivium," the lesser division of study material in the Medieval system, is striking because it is comprised of three subjects—grammar, rhetoric and logic. The number "three" commands interest for many reasons. It is the number of the Godhead—Father, Son and Holy Spirit. It is the foundational number of our Western harmonic system, which is a "triadic" system. In the Hebrew language, the third letter is the "gimel" which is the number of the soul—also, the nurturing aspect or feminine aspect of the alphabet. Understanding comes through language which is comprised of grammar, rhetoric and logic. Philosophically, that makes good sense. To the engineering mind, it may seem a stretch, to say the least. Regardless, however, of the view one takes, without knowledge of the "trivium," all other studies would falter.

The entire concept of the "trivium" and the quadrivium" is, in and of itself, quite intriguing to me. The division, philosophically, into a group of three and a group of four, with the resulting sum of seven seems undeniably well-designed, and certainly not the result of chance, when one takes into account the tightly woven concept of gematria[117] in the ancient Hebrew language.

Additionally, the idea of sound as the cohesive factor in our universe, which is now being studied in intricate fashion, comes into play in a significant way. The first three disciplines (in the trivium) have to do directly

---

117 ge·ma·tria: noun

1: a cryptograph in the form of a word whose letters have the numerical values of a word taken as the hidden meaning

2: the cabalistic method of explaining the Hebrew Scriptures by means of the cryptographic significance of the words; http://unabridged.merriam-webster.com/unabridged/gematria.

with the accompanying idea of sound since they all have to do with the structure and development of language—grammar, rhetoric and the next "logical" step, which is the study of logic[118]—"the science that studies the formal processes used in thinking and reasoning—and, thence—**speaking**." All are facets of sound—vibration, frequency.

Extending this line of thinking to the quadrivium, aided by the sanctified imagination, we might just be able to envision the beginning of **all knowledge.** The four uppermost disciplines in the quadrivium are, as stated earlier, music, arithmetic, geometry and astronomy. This is certainly a skeletal sketch of the body of knowledge, but remember, there has to be a beginning for everything except life, and it is a matter of logical perception that life is the sole source of additional life. Music embodies art; arithmetic is the primitive source of all higher math; geometry is the sperm-cell of all things having design, order and motion; astronomy is the incentive for all scientific considerations. By reducing "the body of knowledge" to these four primordial categories, it becomes increasingly effortless to clearly see how all things are connected.

Genesis, the 4th chapter,[119] makes an interesting addendum to this line of thinking. It names Jabal as the father of those *"who dwell in tents and keep livestock"*—farmers, shepherds and ranchers. Another son, TubalCain, was *"the instructor of every craftsman in bronze, and iron"*—manufacturers and artisans of material things. Jubal was *"the father of all those who play the harp and flute"*—musicians and artists of other genre by inference.

It is of dire importance to be able to see the link between these early and primitive establishments of what we now—in the 21st century—look

---

118   http://www.merriam-webster.com/dictionary/logic.

119   Genesis 4:19-22: NKJV: 1990.

upon as if they had always been. They have not always been; they were created, and the strong inference is of a Divine Creator Who could and did see beyond the time of creation into the eternity which did, after all belong to Him. This being the case, it becomes of equally dire importance to understand the possibility of the creating force in the Person of the Word of God—or, if you can grasp the concept based on the ancient Hebrew— the Song of God—Jesus Christ.

If one can get beyond the very human resistance to the infallibility of the Word of God, it suddenly becomes very clear that the words of Colossians 1:15–16 have a deeper meaning than the meaning that is most often accepted. Those verses read:

> *"For by Him all things were created that are in heaven and that are on earth, visible and invisible, whether thrones or dominions or principalities or powers. All things were created through Him and for Him. And He is before all things, **and in Him all things consist.**"*

For eons gone by, mankind has one way or the other accepted the idea that Jesus, the Son of God, was part of the creation process, and that some kind of sovereign and eternal power equips Him to be, as well, part of the maintaining power which holds the universes together. But a sweeping change has taken place in our body of knowledge. We are now being told that the cohesive force that sustains our entire universe and all that is in it, is **sound**—vibration and frequency. It is of great interest to me how words are used and what veiled inferences there are in the selection of words for any given venue. One of the minute particles of energy that are named as those cohesive forces are, of all things, "**gluons**." Interesting!

Here is the captivating centrality: the speaking voice is activated by vibrating energy and the singing voice is even more energetic than the speaking voice. Therefore, how intriguing to consider the phrase, *"In Him all things consist."* If God Almighty, in the Person of and with the assistance of God the Son, brought *"all things"* into being out of nothing but the energy and power of His voice, how magnificently grand is the idea that the same eternally energetic vibration or frequency of the voice of God—the Word or Song of God—actually **is** the force by which *"all things consist."* This, without any further equivocation, establishes Christ as the centrality, the basis of, the **truth** that He—the Word of God, or concomitantly the Song of God—is **the Central Figure, the Central Force, the absolutely Central Icon of the entire creation saga**. John 1:3 states:

> *"All things were made through Him, and without Him nothing was made that was made."*

This realization immediately establishes the necessity to deal in one's mind and spirit with the Trinitarian God. God the Father, Son and Holy Ghost becomes a construct with which we favored creatures on the face of this planet called earth must deal spiritually and intellectually if we are to be able to grasp the beauty and transforming supremacy of this realization. Within the confines of this realization lies the ultimate solution to dilemmas and unanswered questions which have plagued thinking scholars for eons of time.

The standing problem is the dilemma created by the acceptance of the finite nature of man's mind and intellect. Whether or not one is willing to accept the idea that man is not omniscient or infinitely wise in all realms, one can never solve that dilemma of man's limitations intellectually or spiritually. We can only **know** based on those limitations.

Once that acceptance of limitation is accomplished, the rest becomes a matter of timely learning at the hand of Almighty God—and He has made provision for that learning and given many of His creatures great depth of discernment over the ages.

Consider the New Trivium.

> **Theology**: The themes of theology include God, humanity, the world, salvation and eschatology (the study of last times). The concept of theology that is applicable as a science in all religions and that is therefore neutral is difficult to distill and determine.

It is not necessary to even peripherally address eschatology in this writing at this point. And, actually, it may not be necessary at all, because eschatology is far more a theme of apologetics than it is the theme of this book. Salvation is, of course, a major consideration, because without understanding that theme, accepting other necessary themes almost automatically becomes problematic. But this book is not directed specifically toward the understanding of salvation. Suffice it to say that if Jesus Christ is to be the centrality of the creation saga, His right to that position has everything to do with Who He is and why He is Whom He claims to be. But my sense of balance and acuity dictates that this is too demanding and too spiritually emphatic a theme to deal with here.

The main point of departure between theology and science seems to be at the point of definition itself. By that I mean that theology is defined as the study of "God, humanity, the world, salvation,"[120] and therefore, by definition, it defies the ability to be proven, and embraces almost entirely the issue

---

120   http://www.britannica.com/topic/theology.

of faith, which, of course, cannot be a proven entity and really cannot be defined. It is supremely subjective and involves an attitude of heart which will differ from person to person. Science, on the other hand, is defined as "any system of knowledge that is concerned with the physical world and its phenomena and that entails unbiased observations and systematic experimentation. **In general, a science involves a pursuit of knowledge covering general truths or the operations of fundamental laws.**"[121]

The last statement in the *Britannica* definition of science is critically important: it states that the study of science of any type or genre "involves a pursuit of knowledge covering general **"truths or the operations of fundamental laws."** It may require an open-minded exercise of sanctified imagination to follow the "creative path" to grasping the intense amount of agreement and connectedness that exists among the many genres of science as a whole. Scientists can probably agree that there is, as a matter of fact, a corollary within each various branch of science at the fundamental level where the governing laws are in existence. To pursue that operational corollary becomes vital to the understanding of the idea that the connectedness of all scientific studies infers intense similarity in form, structure and the laws controlling the whole. This may be the point at which some degree of both agreement and disagreement can be evidenced, at least philosophically, and that the resulting agreement or disagreement may allow for a creative path of resolution. To find a path to resolution in the scientific/theological worlds would also be very exciting; I personally see no absolute need for irresolution and conflict on the sub-structural levels.

Personally, as I have pursued my own path of discovery and knowledge, I have come to the place where it is not difficult to accept the tenets

---

121 http://www.britannica.com/topic/science.

of both theology and science as being quite agreeable—one with the other. Human mental acuity in the field of scientific research has become accustomed to demanding empirical proof before acceptance can be acquired. Human mental acuity is, however and finally, finite. If the Creator is, as theologians believe, the exquisitely exclusive and Almighty God, the human mind cannot even approach that level of intellectual and spiritual function. If this Almighty Creator God did, in fact, foster and fashion all of the laws of every discipline, then there is bound to be a high degree of connection and similarity in all created things and entities. The "rub" lies in the matter of **faith**, which is the defining facet of theology. But **faith** is not a foreign idea in the world of science either. Einstein spoke about faith often, and Max Planck probably made the most intriguing statements about that idea in the scientific realm when he said:

> "Anybody who has been seriously engaged in scientific work of any kind realizes that over the entrance to the gates of the temple of science are written the words: 'Ye must have faith.'"

For the scientific mind to embark on any new idea or quest, there must be an element of "faith"—faith in the idea as being violable, faith in the ultimate result of the research, faith in the precursor of the idea to be able to follow it to maturity. But "faith" is a must. The great thinkers of the early ages of scientific pursuit knew that, and many have written quite eloquently about that "faith." Some have even related it to some degree or other to the faith that one would have in God. The questions still prevail, but the honest scientist recognizes the element of "faith," and does not accede to the idea that it does not exist or that it has no bearing in the empirical world of science.

Faith, like music,[122] **seems to have no definable beginnings**. It can be elaborated upon, but it has no real genesis that we can find. It is a personal belief of mine, backed by my entire theory, that this is true of more of the cosmos than we have taken bother to investigate. "Faith *is*." And as such, is applicable to many venues, but is beyond human invention. "Music *is*." And it enjoys the same eternality, if you please, that faith encompasses. And it would be an incredibly fascinating study to pursue the geneses of many other things that we take for granted and to discover that they have the same "forever" qualities. But that is for another time, in another venue.

When dealing with the idea of creation, it is safe to say that most of the Judeo/Christian world is agreed that God's creation was complete and fully functional from the start. But how many of the great scholars and thinkers of the "Christian faith" have taken their thinking to the deepest level and made the correct assumption that the Word of God, Jesus Christ, cannot reasonably be seen as only a tangential force in the entire creation saga? Again, I assert that this book is not an apologetic for the Christian faith, but this belief, which harks back to the first assertions of the Book of Genesis, is the bedrock of what we call "faith" to begin with, and must be dealt with as such. In 1911, Jan H. Boer translated a statement by Abraham Kuyper, the great Dutch theologian, university professor and prime-minister:

> "The Son (Jesus Christ) is not to be excluded from any-
> thing. You cannot point to any natural realm or star or
> comet or even descend into the depth of the earth, but it
> is related to Christ, not in some tangential, unimportant

---

122 Curt Sachs (born June 29, 1881, Berlin, Germany—died Feb. 5, 1959, New York, N.Y., U.S.), eminent German musicologist, teacher, and authority on musical instruments, said, "However far back
we trace mankind, we fail to see the springing up of music. Even the most primitive tribes are musically beyond the first attempts."

way but directly. There is no force in nature, no laws that control those forces that do not have their origin in that eternal Word. For this reason, it is totally false to restrict Christ to spiritual affairs and to assert that there is no point of contact between Him and the natural sciences."[123]

In spite of these strong words, how many of us can claim that we were taught in our youth that there was a centrality to creation and that there was a connection and powerful pervasiveness of "simple and beautiful laws" undergirding everything that we know? Even when I studied about Einstein in later years, it was not made clear that this notion was his motivating and energizing conviction.

Because its basic historical roots are in Judaism, "faith" has become a consideration of narrow import—mostly in the world of religion. It can be extended to other religions because it deals with the relationships of man with whatever gods or God may be extant in a particular culture, but its critical considerations are, because of the involvement of Jesus Christ with mankind, Christian and pertain to the "Christian faith." But it is interesting to suppose that the concept of faith as set forth by the Christian belief has had the powerful effect on thinking in all realms, simply by virtue of its nature.

"Faith" can be viewed in eternal terms historically. How it has proliferated into other formats is quite amazing. Simply put, **"In religious traditions stressing divine grace, it is the inner certainty or attitude of love granted by God Himself. In Christian theology, faith is the divinely inspired human response to God's historical revelation**

---

123 Abraham Kuyper, *You Can Do Greater Things Than Christ*, Kamper, The Netherlands, J. H. Kok, 1911, Viewed at Google Books.

through Jesus Christ and, consequently, is of crucial significance."[124]
The matter of "faith" in the Divine sense includes the idea of "complete-ness," "irrevocability" and "perfection"—the essence of the things in which we have "**faith**."

Genesis 1:31 reads: *"God saw everything that He had made, and indeed it was very good."*

A cursory word study on the key words of this verse reveals a wealth of information which cannot be found in the simple English translation of it. The nature of the Hebrew language provides a rich backdrop for deep interpretations of the original Scriptures, which the English language does not. The word "saw," in the Hebrew, contains the ideas of having "a vision for, to learn about, to discern" and even more. The inference could be, if viewed by "faith," that God's total reaction and understanding of what He had done—was "good." That word, "good," is an all-encompassing word in the original language.[125] It has intellectual, physical, emotional and visionary qualities, and also affirms the gamut of understanding of the matter of faith.

It is all but impossible to consider the concept of "theology" with-out first considering the concept of "faith." Faith facilitates the ideology of theology. Without an understanding of faith, the core of theology is annihilated. But the interesting bit of connection between these two disci-plines—science and theology—which so often are at war, is that the matter of faith is so prevalent in both. Why, then, the dichotomy? If I dare, I would say that to have science demanding empirical evidence when there is so much evidence of the element of faith in science itself is an oddity.

124 Ibid. Pg. 75.

125 https://www.blueletterbible.org/nkjv/gen/1/31/t_conc_1031.

Perhaps, if this transient "half-idea" that theology and science have, at least, some of the same ingredients could be further developed, the great gulf between could be shrunk. It is a matter of "sanctified imagination" to be able to allow one's self to see the possibility.

It is well-stated in the *Britannica* definition of **theology**: "The concept of theology that is applicable as a science in all religions and that is therefore neutral is difficult to distill and determine."

"**Science** (is) any system of knowledge that is concerned with the physical world and its phenomena and that entails unbiased observations and systematic experimentation. In general, a science involves a pursuit of knowledge covering general truths or the operation of fundamental laws."

While it is difficult to actually perceive theology as a science and see it as means to unbiased observation and systematic experimentation, I believe that there is a very clear sense in which science and theology can share open space together and even become mutually inclusive.

Britannica astutely defines science in general as a "pursuit of knowledge covering general truths or the operation of fundamental laws." This assumption in no way contradicts or defies the "general truths" or "fundamental laws" that underlie the "system of knowledge" that we call theology. The problem lies in the perception of theology as a "philosophically oriented discipline of religious speculation and apologetics that is traditionally restricted, because of its origins and format, to Christianity...."[126] The themes of theology—"God, humanity, the world, salvation, and eschatology (the study of last times)"—further complicate sorting out the differences in perception, because, characteristically, such subject

---

126 Ibid pg. 185: *God's Opera*.

matter is quite foreign to our human understanding and there may even be a significant amount of reserve or resistance to cognitive acceptance of it.

Christopher Hitchens, English author, journalist, religious and literary critic, and well-known atheist, made the observation that, if, indeed, there were a God, it would be incumbent upon mankind and even mandatory for mankind to bend his will to that God. The outcome of that mandate would, of course, be preposterous and therefore, completely illogical. This thinking fosters all sorts of speculation and unbelief, and after generations and eons of time, this line of thought can create a mindset that belies the possibility of seeing things spiritual as nothing more than philosophical meanderings. What I am hoping to see is another shift of paradigm as thoughtful, objective thinkers begin to glimpse the possibilities of aligning science and theology as fellow-disciplines with similar worth and similar intellectual integrity. There is an essential facet to this shift of paradigm: one must see the possibility of God in the first place and also the possibility of that God's being at the center of all that is, instead of seeing the intellect of human kind at the center of all that is.

That shift would enable the general truths and fundamental laws of both disciplines to be seen as coherent and consistent—and connected. It would probably become necessary under the scrutiny of that shift of paradigm, to examine the associated possibility that all disciplines could be tested for validity and worth with the same scientific methods that now control the existing and accepted sciences—the burgeoning of knowledge after the advent of the Age of Reason—the enlightenment—necessitated the ushering in of the age of specialization as well. So much knowledge growing at such a fast pace necessitated a polarization of that knowledge so that it could be properly evaluated and categorized. It does not surprise me that very soon after mankind began to "reason" and to develop volumes of writings on volumes of subjects, the need arose to partition that vast learning into manageable categories and therefore to give the various fields a varying set of guiding laws and guidelines. This inevitably led to the fracturing of the connection, and the segregation of each separate discipline from that connectedness created an excess of individual volumes of knowledge which seemed to be moving further and further away from the center of common knowledge of common subjects. Mathematics could no longer be considered with science or art; it had to become its own self with its own ruling factors. Science could no longer be accepted in the same light as theology because scientific information could be experimented with and "proven" while theology, as it proliferated in scope, had to be categorically quantified by faith. It could **not** be proven. Art, of course, could absolutely not be quantified or analyzed along with mathematics, because it was too subjective and too short-lived a happening and depended too much on mood or culture. And soon, it became impossible to see the art in a mathematical equation because of the resultant strict categorization and imposed compartmentalization.

And so it went until there was so much knowledge, so much information, so many subjects that it was "obvious" that they **all** had to have their own identities and their own controlling rules and laws. Connectedness and agreement become less and less plausible.

But remember, **"Science is any system of knowledge that is concerned with the physical world and its phenomena and that entails unbiased observations and systematic experimentation—***(it is) a pursuit of knowledge covering general truths or the operation of fundamental laws.*****"**[127] Therefore, I submit that **all knowledge can be** examined

---

127 Ibid: *God's Opera.*

under the canopy of **science** for the simple reason that, by definition, "science is a (the) pursuit of knowledge covering (the) general truths or the operation of fundamental laws." And I believe that one would be hard pressed to find sufficient proof of the absence of general truths or fundamental laws in all knowledge, if one were to be open-minded and able to use one's sanctified imagination in these considerations.

As a matter of fact, being able to see and understand the underlying relationships of knowledge and discovery as Einstein did, is incredibly exciting, and, in fact, is the very motivation that kept Einstein searching and moving ahead in his quest for over thirty years. At one point in his quest, he tried sincerely to contradict his own theories by re-working his equations, but the strength of his equations would not permit him to depart from the belief that there, indeed, was a consistency in his understanding of general relativity and quantum mechanics that could not be destroyed. And so he renewed his search for that "beautiful and simple set of laws" that could, and eventually, would, explain all of creation, and maintained that search for the rest of his life. His fanatical belief in the strength of imaginative thinking was his preserving ideal.

The one thing that makes it harder to follow that same dogged and unrelenting sort of thinking is the abundance of knowledge that engulfs us in the 21st century.

Art (is) the study of the nature of art, including such concepts as interpretation, representation and expression and form. Perhaps a more understandable definition for the purposes of this book would be the definition of **aesthetics: Aesthetics, the philosophical study of *beauty and taste*. It is closely related to the philosophy of art, which is concerned with the nature of art and the concepts in terms of which individual works of art are interpreted and evaluated.**

170

It is necessary to be careful of semantics in dealing with this kind of study because all of the subject matter is so vast, and narrowing the choice of issues to be dealt with is essential. The evaluation of art (or aesthetics) is, in my estimation, even more difficult than the evaluation of theology, because "truth" in theology is more accessible because of the nature of the Source of theology, God Himself, than "truth" in art.

Robert Shaw came closest to isolating that quest and narrowing it into an intellectually accessible study. He was convinced that "truth in music" was real, quantifiable and necessary to be addressed. At the time that I was singing under his baton, in the late '70's and early '80's, neither he nor any of us 250 or so singers had a very clear idea of what that meant. And maybe most musicians still do not perceive exactly what he was getting at, but here is the fundamental issue: If Einstein were correct, if the Biblical record is sound, if science is getting closer to the central issues of creation and the universe, if there is a connectedness in all creation and if the fingerprint of God Almighty is, in fact, on all creation, then all knowledge will support and point directly to the same "truth," including art—and music.

It is easier to quantify and qualify the elements of music than it is to do the same in other related art-forms. However, I believe that it is already very clear that visual art, for instance, is governed by the same laws of frequency, balance, timbre, rhythm and that those same values can be easily applied to, for instance, sculpture, architecture, dance, poetry. In fact, if one would take the time and leisure to meditate on the qualities of artistic expression, one would have little trouble in seeing the similarities, at least with the imaginative eye of creativity.

God actually, in my sanctified imagination, gave us an extremely valuable and limitless gift when He gave us co-creative skills that enable

humankind to compose music, paint paintings, write poems, build buildings of great beauty, sculpt great statues, tell long and beautiful tales of conquest—there is no end to it and that is precisely what makes it so intriguingly majestic and inexhaustible.

Stop to think: even a non-musician can appreciate this nugget of truth. There are only 12 semi-tones in the musical scale which defines our Western musical theory framework. Every 13th tone, the sound repeats itself at the octave, which is a direct proof of a mathematical truth which delineates and defines the number of vibrations per second that creates that sound— ad infinitum, until the human ear can no longer discern the sound. Out of 12 semi-tones comes all Bach, Beethoven, Brahms, Chopin, Grainger, Bernstein, Mozart, Haydn, Rodgers, Hammerstein—on and on. Out of 12 semi-tones comes all piano, guitar, organ, orchestral, trumpet, violin, cello music—on and on. This is, in and of itself, miraculous and a very plausible example of the vastness of all nature which God has created: no fingerprints are identical, no DNA is identical, no snowflakes are identical.

A New Trivium?

Allow me to propose a thought: What if we took these three fields of study, and approached them all as sciences? What if we could allow ourselves to look at everything we know as if it were viable and had the possibility of being evaluated and assessed using the empirical processes of scientific research as we define it today? I believe it possible that we could discover new "truth" and establish new continuity in our thinking and new, exciting views of the connection in our bodies of knowledge.

Scene II

FAITH DESCRIBED—IT DOES NOT BELONG TO THEOLOGY ALONE

> Consider this…
>
> *Scientists declare that "faith" of some sort or another is inherent to the practice of good science. The funny thing is, this is exactly what Christians have said for centuries.*

Anybody who has been seriously engaged in scientific work of any kind realizes that over the entrance to the gates of the temple of science are written the words: 'Ye must have faith.'

All matter originates and exists only by virtue of a force… We must assume behind this force the existence of a conscious and intelligent Mind. This Mind is the matrix of all matter.

Whence come I and whither go I? That is the great unfathomable question, the same for every one of us. Science has no answer to it.

Max Planck[128]

Imagination is more important than knowledge.

Albert Einstein[129]

128   http://www.brainyquote.com/quotes/authors/m/max_planck.html.

129   http://rescomp.stanford.edu/~cheshire/EinsteinQuotes.html.

I t is a very slippery slope. When dealing with the world of theology, it soon becomes evident that there is a gulf quite well-fixed between the philosophy of the conservative and the philosophy of the liberal. And, furthermore, that gulf widens when the conversation turns to things scientific.

Over the last century or so, that gulf has become almost too wide to navigate. In loose terms, the scientist will declare that the "scientific method" demands empirical proof, while, at the same time, he/she will also declare that the very nature of theology makes it impossible to provide that proof. It is puzzling that, at the same time, he/she or someone of his/her colleagues will also declare that faith of some sort or other is inherent to the practice of good science.

The common residence for the word "faith" is in the documents and sermons of the theologian as he/she writes and propounds the huge library of thoughts and information about God and man, and the interactions between them. Along these lines, I must remind the reader that I have one-among-many foundational precepts that governs my thinking, and it is this:

> God wastes nothing. When He gave mankind the precept
> of "faith," He made that principal operational in every
> way and place possible. It is a necessary component in
> life in the broadest sense. Everything is for our edification
> and education and the connections are myriad.

But, think about this: how often does the common man stop to consider that faith is an undeniable factor in medicine, in engineering, in business? In education? In relationships of all kinds? We would have no banks, no mail delivery, no retail establishments, no schools if it were not for the concept of faith. Even language is influenced by the idea of faith in something or somebody. We are bound by the laws and constraints of faith from very deep in our psyches. And this is expressed quite clearly from time to time—**but**—we bridle our understandings and imaginations and don't pay the right kind of attention: the very word "credit" comes from an ancient language word root—"credo"—which means simply "I believe."

The evolution of the word "credit" is a direct affirmation of the fact that faith is intrinsic to our entire culture. When an institution extends "credit" to us, they are exhibiting their "faith" in us to pay back, with interest, probably, the entire amount of credit extended. When a citizen pays for something with his "credit" card, the payee is exhibiting "faith" in the institution holding the card's debts to pay that obligation, and the holding institution is exhibiting "faith" in the card holder to pay off that debt. Interesting. Faith and credit intricately entwined.

OVER THE ENTRANCE TO THE GATE OF THE TEMPLE OF SCIENCE ARE WRITTEN THE WORDS: 'YE MUST HAVE FAITH'

"What is faith? It is confidence that someone or something is reliable. Our whole life is based on faith…Believing (faith) is intellectual in the sense that faith has to have some facts to rest upon."[130]

If you consider the accomplishments of any famous scientist, you can easily see that, first, there was an idea—a notion about something. And then, there was arduous work filled with success **and** failure. Each new achievement is built on the successes or failures of the ones that preceded

---

130 Word Study: *The New Open Bible*: Pg. 1461: copyright 1997 by Dr. D. James Kennedy. All rights reserved.

it. But always, there is the element of faith—faith in the ideas **and** realizations of those ideas which were brought to fruition by the "believing" scholars and workers who saw far ahead into the future—and who became the flagship captains of any one of the myriads of new miracles in the fields of endeavor involved.

We and our parents and grandparents have lived in a century that saw humankind advance from riding horseback between Washington, D.C., and Richmond, Virginia, for example, to do the work of government, to Presidents who traverse the globe in Air Force One at over 600 MPH. It took an immense amount of faith for Wilbur and Orville Wright to design, build and climb aboard that flimsy wooden "model airplane" and allow it to take them 100 feet into the air—only to crash land and then try again. In less than 100 years, life has forever been transformed by the faith of those two brothers who responded to an idea that catapulted out of their sanctified imaginations, gave it wings, so to speak, and had the faith to pursue their dream. But, consider—their faith was not new energy put to task—hundreds of years before, Leonardo da Vinci had conceptualized flight and put that idea on paper. Faith in his calculations must have been part of Wilbur's and Orville's thinking.

The problem is one of appropriation. Recently, I have been engaging in interesting conversations with a priest friend of mine about faith and the divergence between conservatives and liberals in the world of theology. Territorialism is very evident in our theological communities in the 21st century, and there seems to be a latent sort of pride with which we guard our own small personal beliefs. Our own particular faith system seems to be one of the things heavily walled-off by our protective shell. By that, I mean the liberal often speaks of himself as being a "flaming liberal," and the inference is that, since fire burns, it might be better not to challenge this system too far.

On the other hand, the conservative seems, sometimes, to be able to arm himself with the "mind of Christ" which, thereby, gives him the right to be staunchly inflexible in his beliefs for the precise reason that he has that mind of Deity which is infallible. Both could be desperately wrong. One of the things that first caught my attention when I began thinking thoughts straight out of my more unconventional, outside-the-box mind was the idea that the enlightenment was not necessarily a wondrously good thing that happened during the course of history. It is true and easily acceptable to understand that without the Age of Enlightenment, freedoms of all kinds would have gone begging, knowledge would have foundered before it had time to flower, human rights of all sorts would have continued in a state of scarcity; progress would have been shackled.

In the first place, simply by virtue of being an evangelical or of a conservative religious mindset, this does not give one the obvious right to claim the knowledge of truth and reason without reservation. And just as well, the liberal mind really does not have the right to assume that all wisdom, all intelligence, all consuming ability to reason comes automatically with the right to unfettered thought about human supremacy. But, to me, one of the very obvious fallout results of the Age of Enlightenment is "entitlement." And that is an attitude that is so very easy to assimilate.

Another one of the outside-the-box revelations that I had early on was the idea of human supremacy and how it conflicted with my religious upbringing. I noticed early on in my college years, which started in my forties, that the general and largely unspoken attitude was that the human mind and intellect was, in fact, supreme. No matter what the subject of study, the inherent, latent underlying intelligence was human in nature. Man had the preeminence, certainly. I balked at the idea. I balked because I was a conservative and totally of a conservative theo-

logical mindset—I was called a "fundamentalist." I had that point of view so deeply ingrained in my subconscious that, indeed, I could go as far as to say that man was unworthy, decadent and sinful from the start. And so to elevate him to this degree had to be evil. I hope it is becoming more acceptable to see that neither response is completely valid. Man has a God-granted place of preeminence over other animal forms, but he is not inherently supreme, nor is he inherently at the center of all creation. He is, in fact, in need of redemption from the Holy Spirit point of view. My theological sensibilities require for me to understand atonement. But again that is not the subject of this writing; it is an essential ingredient in understanding.

The result of seeing man as the center of all creation is, that by the 21st century, man has become more or less completely convinced, as a generic group, that he is entitled, intelligent to the utmost, to be the center of all knowledge in this universe. He is the creator of music, he is the creator of knowledge, it is through his pipeline that all is accomplished. In some circles, there is little or no need for God.

Of course, this is a philosophical discussion that is far, far beyond the parameters of this book, and, once more, I must offer a caveat: it is not my intent to write a book on apologetics. I am simply trying to dig up the soil where a new and more beautifully prolific garden of ideas can flourish. However, it is mandatory for the purposes of this book to present the challenge to explore possibilities and to give way to new insights.

Our entire universe is filled with examples of the need for faith: the problem is to discover what faith **really is:** "Look deep into nature, and then you will understand everything better." It is evident in this statement that Einstein had the ability to think intuitively. He was not an outspoken conservative or believer, but his writings are complex and when taken in their entirety, point to an equally complex belief in the existence of God. This is a quote from his creed, titled "What I Believe," written in 1930:

> "To sense that behind everything that can be experienced there is something that our minds cannot grasp, whose beauty and sublimity reaches us only indirectly: this is religiousness. In this sense...I am a devoutly religious man."[131]

In response to a young girl who had asked him whether he believed in God, he wrote: "Everyone who is seriously involved in the pursuit of science becomes convinced that a spirit is manifest in the laws of the universe—a spirit vastly superior to that of man." And during a talk at Union Theological Seminary on the relationship between religion and science, Einstein declared: "The situation may be expressed by an image. Science without religion is lame; religion without science is blind."

While I unabashedly take the position that Jesus Christ is the incarnate Word of God, and that statement by itself infers that the core Source of sound and the scientifically-proven functions of sound are in Christ, I still hold to the immediate necessity to listen to and consider the intuitive statements of men like Einstein, Planck, Newton and Pythagoras as they wrestled with the same veiled **references to** and **results of** the presence of the Divine Creator—Sustainer of all life—now, and from the first beginnings, and on into the unseen halls of eternity. Max Planck had a veiled insight into the need for recognizing and accepting the omnipotent and omniscient presence of God the Creator—even if he did not name Him, specifically. What a marvelous indication of the validity of consulting with one's intuitive, sanctified imagination. Take note of Planck's insight:

131 http://www.strangenotions.com/einstein-god/.

"ALL MATTER ORIGINATES AND EXISTS ONLY BY VIRTUE OF A FORCE..."

Planck goes on to expand upon this provocative statement:

> "We must assume behind this force the existence of a conscious and intelligent Mind. This Mind is the matrix of all matter."[132]

In the final analysis, we who grapple with this question of origins, whether origins of music, where I started my quest, or origins of language or the final essential origins—mankind and all that is embraced by the vast panoply of knowledge that accompanies this subject—must make one mammoth decision: where is our basic belief system rooted? In the supremacy of man and his world or in the over-arching supremacy of God and His inexplicable presence in all things? If we can cope with this one idea and come to the place where we are at least willing to consider God as central to all knowledge, all understanding and all being of all kinds, **and all redemption**, then we can begin the journey to what I dare to call real faith—faith in God Himself. All else is simply fallout from that incredible truth—God Is.

Of course, that also means that we must begin to consider that God, as He describes Himself, is also Truth, and, then, that realization embodies the Trinitarian concept of the Holy One. Then, issue by issue, the mysteries of life become less mysterious. Even the deepest issues begin to appear more closely woven into a beautiful pattern which only ratifies the presence and orderly orchestration of the Divine. There is a vast difference between the harsh-sounding tuning of the orchestra and the indescribable beauty of,

132   Ibid.

say, a Brahms' symphony. The house (universe) without a foundation is doomed to collapse. Science does finally understand that "all matter originates and exists only by virtue of a force." How necessary to have a strong underlying and eloquent foundation which will support the full meaning of assuming that "behind this force (is) the existence of a conscious and intelligent Mind. This Mind is the matrix of all matter." How necessary to at least **consider**, at this point, that the centrality **is** God Triune.

WHENCE COME I AND WHITHER GO I? THAT IS THE GREAT UNFATHOMABLE QUESTION, THE SAME FOR EVERY ONE OF US. SCIENCE HAS NO ANSWER TO IT.

"Science cannot solve the ultimate mystery of nature. And that is because, in the last analysis, we ourselves are a part of the mystery that we are trying to solve."[133] Max Planck.

Although my world view has always been that of the conservative and evangelical bent, I am still convinced that in order to know the mind of God, as the Scripture tells us we must do, we must be willing to engage our intuitive and sanctified imaginations in order to move outside of our own narrowness. Hence, my cherished relationship with my inner alter-ego, Jana, who has always had the capacity to imagine the most beautiful, the most exquisite imaginings and under the tutorship of many marvelous spiritual mentors has afforded me invaluable riches of thought and learning.

My inclination to disavow the all-inclusive, ubiquitous worth of what we call the Age of Enlightenment can be problematic. Coming into that place of intellectual expansion was, indeed, a very necessary component in the ages-long evolution of mankind.[134] The ensuing explosion of

---

133 http://www.brainyquote.com/quotes/authors/m/max_planck.html#4BJ04QlSub8P2iYM.99.

134 Please understand that I am **not** referring to Darwinian evolution, but, rather, the overall advancement of all mankind and his vast repertoire of accomplishments throughout the ages.

knowledge and increased benefits to all mankind and all of creation is a matter of great importance. It had to happen in exactly the way that it did.

The whole concept of inevitability comes into play here. Einstien grappled with "inevitability" in physics. Mathematicians know the dilemma of "inevitability" when dealing with numbers and equations. Engineers understand the "inevitability" of proportion and equation as well. As I ponder these ideas, I can see more clearly the necessity—the "inevitability," if you will—of recognizing the necessity for acknowledging the presence of God the Creator in all things—simply by the preponderance of evidence—the connection and association that is so evident in all of creation demands that acknowledgement. That irrefutable design, which predominates the entire universe, demands a Supreme Designer.

I believe that most scientists, philosophers and thinkers from any and all fields of knowledge have little difficulty with seeing the evidence for, and maybe even the need for, a Grand Designer; the great schism is in conceding that this Grand Designer is, in fact, Almighty God, Who presents Himself as Master and Creator of all and can be known only as a result of a personal one-on-one relationship with Him. That is a heavy burden for proud mankind to subject himself to, because it involves man's moral weakness, as well as the concepts of eternity and omniscience, which the common man may have trouble understanding and acquiescing to.

Once the chasm of that acquiescence is bridged, all else becomes much more palatable and much more easily assimilated. That acceptance of the Almighty has two dimensions: it is broad and it is specific, at one and the same time. The broadness of that acceptance is the easier of the two, and I believe that most scientists who move and breathe in the realm of the marvels of life and the universe that supports that life have less trouble with this dimension. Of course, there must be intelligent design.

We can see it everywhere, and its presence can even be a comfort in our diverse, multifaceted and demanding world. It brings reason and order, which otherwise could not be even minimally described in reference to our world and the universe at large. It also seems reasonable that the more deeply we go into the exploration of our universe and the life that it supports, the more easily accessible the understanding of that reason and order should become. But mysteriously, that is not the case.

The specificity of the acceptance of the truth of the Almighty God as elemental and irrefutable is quite another concept to be dealt with. The question of origins and creation always carries with it the quandary: *"Whence come I, and whither go I?"* And thus far, some of the greatest of scholars and thinkers have failed to find the answer, because, I think, to find that answer requires that we also accept the unquestionable Sovereignty of that God of Whom we speak. We are required to then put Him in the place of sovereignty, which means that we accept Him as the ultimate Power behind every event and as the ultimate Answer to every need, dilemma or condition that we as His creatures may experience. And man, the "created being," must become subordinate in the order of things.

"And therein lies the rub." To see humanity as not being the intellectual and studied center of **all** is difficult, since it relates rather directly to the evolution of the Age of Enlightenment and all that is consequent to that evolution. The quote from Max Planck says it so very well: "Science cannot solve the ultimate mystery of nature. And that is because, in the last analysis, we ourselves are a part of the mystery that we are trying to solve."

IMAGINATION IS MORE IMPORTANT THAN KNOWLEDGE

What if there is a Supreme Being? What if that Supreme Being does have **all power, all knowledge, all creative ability** over and far beyond anything which the human mind can comprehend? Could it be possible

that there is understanding and knowledge far beyond our human, finite minds' capability to grasp? Is it imperative to circumvent these possibilities in order to preserve the grandiose abilities of the human animal?

Imagine the worlds of information that may be out there, somewhere, just waiting for the sanctified imagination of a few daring human beings who are willing to subordinate their ultimate egoistic selves to the "unknown" God. The Romans had great difficulty with this notion. It openly conflicted with their concept of the Caesars as gods, and they were reluctant to abandon that concept.

Over the eons, little has changed. Man is still not willing to accept being subordinate to a Higher Being Whom he cannot even see. However, it is interesting that the greatest minds have, down through the ages, made that necessary acquiescence to some degree or other that gives them a glimpse of the necessity for realizing that there must be a power beyond the human vista. Many have brushed shoulders with the full knowledge of that God as Creator, Savior and complete omniscience without quite making the complete journey to understanding in terms of their own finite and human limitations. Let's follow the path of this narrative to see if, how and when it will take the essential turn to the path of truth.

Scene III

THEOLOGY—AN ANCIENT DISCIPLINE REVISITED

> *Consider this…*
>
> ### Let's leave Plato's "polemical intentions" and simply accept the study of theology as the study of God Almighty.

"The concept of theology that is applicable as a science in all religions and that is therefore neutral is difficult to distill and determine. The problem lies in the fact that, whereas theology as a concept had its origins in the tradition of the ancient Greeks, it obtained its content and method only within Christianity. Thus, theology, because of its peculiarly Christian profile, is not readily transferable in its narrow sense to any other religion. In its broader thematic concerns, however, theology as a subject matter is germane to other religions.

"The Greek philosopher Plato, with whom the concept emerges for the first time, associated with the term theology a polemical intention—

"In spite of all the contradictions and nuances that were to emerge in the understanding of this concept in various Christian confessions and schools of thought, a formal criterion remains constant: theology is the attempt of adherents of a faith to represent their statements of belief consistently, to explicate them out of the basis (or fundamen-

185

tals) of their faith, and to assign to such statements their specific place within the context of all other worldly relations (e.g., nature and history) and spiritual processes (e.g., reason and logic).

*"Here, then, the above indicated difficulty becomes apparent. In the first place, theology is a spiritual or religious attempt of "believers" to explicate their faith. In this sense **it is not neutral** and is not attempted from the perspective of removed observation...."[135]*

he chasm that exists between theology and science is not a great mystery. Science, at least as it is understood today, demands neutrality and objectivity. Theology, on the other hand, is dictated to by myriads of differing philosophies and inquiries into subject matter which has more to do with the spirit than it does with the mind. Science is empirical in nature: theology is conjectural in nature because it has to do with vastly different religions and cultures.

That is not to say however, that the two disciplines are impossibly confrontational. Take note:

> "...theology is the attempt of adherents of a faith to represent their statements of belief consistently, to explicate them out of the basis (or fundamentals) of their faith, and to assign to such statements their specific place within the context of all other worldly relations (e.g., nature and history) and spiritual processes (e.g., reason and logic)."[136]

135 http://www.britannica.com/topic/theology.

136 Ibid: *God's Opera*: Above.

Act 3 — Scene III

One of the most obvious of disparities is the overwhelming number of religions and faiths that demand addressing and the fact that each one demands audience in its own right. The same may be said of science because that discipline is now moving in so many directions at once, and each must be empirically addressed in its own right.

The demand for neutrality in science and the absence of neutrality in theology is exacerbated by the fact that the perspective of science is "removed observation," while the perspective of theology is the logical absence of "removed observation."

Overriding all of these perspectives are the gnawing limitations of human intellect. In all actuality, we humans are limited by the very boundaries of our longevity, and all of our most scholarly voyages into the world of knowledge are finite. And, therefore, changeable and vacillating.

The very roots of theology express this inadequacy:

> "The Greek philosopher Plato, with whom the concept emerges for the first time, associated 'a polemical intention' with the term theology."[137]

Polemical means controversial or being against an opinion or doctrine. "Polemical intention" can only beget "polemical intention."

Therefore, the creative path to real understanding of the possibilities of some agreeable ground's being established lies in the opening, once more, of our sanctified imaginations and our thoughtful "alter egos," in order to explore new and fascinating relationships in an ancient discipline with which mankind feels most familiar. The dogged determina-

---

137   Ibid: *God's Opera*: Above.

tion to think of man and his enlightenments as the ultimate authority may be detrimental.

The original "Age of Enlightenment" may have been in the Garden of Eden. It was not the act of disobedience embodied by eating the fruit of the forbidden Tree of Life that brought down God's anger: it was the act of pride that caused the first woman to respond by wanting to be like God and know good from evil. This is what Satan taunted Eve with: to be like God and to know all. Fact or fiction? What if it is, after all, true?

Let's leave Plato's "polemical intentions" for scholars of other schools, and simply accept the study of theology as the study of God Almighty for the purpose of this writing. Revolutionary? I hope so.

Think about it—

Scene IV

THE SCIENCE OF THEOLOGY—THE THEOLOGY OF SCIENCE

> Consider this…
>
> **The basic idea to convey to you is the idea of relationships or "relatedness."**

Science is any "system of knowledge that is concerned with the physical world and its phenomena and that entails unbiased observations and systematic experimentation. In general, a science involves a pursuit of knowledge covering general truths or the operations of fundamental laws."[138]

Theology also deals with the physical world and its phenomena. The first book of the Old Testament is the very story of the creation of that world and the phenomenal beginnings of all things. The most astounding of all is the beginning of the energy-producing star on which we live. But think along with your personal imagination beyond that star—the earth—to the entire universe. Converse with your inner thoughts and begin to see the wonder of all that is—and you will soon begin to understand that all of the systems that converge to support us and our place in this universe were also accomplished in perfect working order at the moment that creation took place. Nothing was left out.

138 http://www.britannica.com/topic/science.

It is magnificently interesting to allow oneself—courtesy of one's own internal thought—to realize that the first law of thermodynamics, which underlies so much of our knowledge of physics, was put into place at that time. That law states that "the total energy in a system is constant. It can change form but cannot be created and cannot be destroyed."[139]

This is science—pure and simple—and it comes directly out of the theological writings of the first book of the Bible—Genesis. The ongoing difficulty is the "enlightened" state of man's understanding without the overarching guidance of the Holy Spirit, which convinces man that he is the originator and center of knowledge. That immediately erases the clear vision of God as Creator and Director of this universe.

Only a truly "scientific appraisal" of these general truths and fundamental laws will allow the unadulterated truth to come to the fore.

This is the "Science of Theology." Could it also be the "Theology of Science"?

IN GENERAL, A SCIENCE INVOLVES A PURSUIT OF KNOWLEDGE COVERING GENERAL TRUTHS OR THE OPERATIONS OF FUNDAMENTAL LAWS.

Given, this is a "general" statement. But, what if in your sanctified imagination, you could begin—since this is a writing about beginnings—to view theology and Scripture as an original expression of beginnings, even the beginnings of systems and science itself? It should not be a foregone conclusion that anything recorded in the Holy Writs is automatically suspect or to be viewed in a more analogous

---

139 *The D. James Kennedy Topical Study Bible*: Charisma Media/Charisma House Book Group: 2015: Pg.6.

190

way than if it were a textbook depicting the orbits of the planets. Scientific methods will work in any sphere if they are indeed unbiased and systematic.

As a matter of fact, by definition, science is a system of knowledge and the investigation of that knowledge, that deals with nature, the stated mechanics of nature and all manner of phenomena with which we come in contact daily. What better investigative system than that to discover the intricacies of the Scriptures, which, by the way, theologians may just miss expressly because they are more attuned to the matters of the spirit and the impressions of the mind, humankind and others. From that vantage point, the marriage of science and theology does not seem at all strange.

I sometimes get stuck on a thought, and that is precisely when I call upon my imagination. Einstein and Planck, among others, knew well that when their scientific natures locked them into an unrelenting pattern of thought, their imaginative natures could, and probably would, rescue them and give them a fresh, new approach.

There can't be a more fertile field of untapped ideas than those found in the Holy Writings. Instead of stubbornly determining to disprove or confound theology, why not put science to work to expose some of the evasive ideas that have been a vague part of our heritage for eons? This includes ideas like creation. Thousands of years of study have not succeeded in solving the enigma of creation. Why not take the Scriptural account and apply scientific methods of study to it with an eye to possibly accepting what we discover?

Science, by definition, is concerned with the physical world and its phenomena; theology, by definition, traces man's interaction with the Being Who conceived of and brought to fruition that physical world and its phenomena. Close enough introspection, I believe, would make this idea very clear and even turn it into a viable theory. If we are not too

enlightened to allow the process. Remember—science is a discipline almost totally dependent upon "processes."

> "The concept of theology that is applicable as a science in all religions and that is therefore neutral is difficult to distill and determine. The problem lies in the fact that, whereas theology as a concept had its origins in the tradition of the ancient Greeks, it obtained its content and method only within Christianity. Thus, theology, because of its peculiarly Christian profile, is not readily transferable in its narrow sense to any other religion. In its broader thematic concerns, however, theology as a subject matter is germane to other religions."[140]

It is interesting to me that *Encyclopedia Britannica* makes this inference: **"The concept of theology that is applicable as a science in all religions and that is therefore neutral is difficult to distill and determine."** This idea is not clarified, as far as I could tell, in the latter parts of the article. In fact, it was startling to see that such a well-documented research publication would state that theology could be applicable as a science. This is quite contrary to my own experience with Christian dogma. But, it opens a door for some profitable conjecture. There actually does not seem to be a huge difference between science and theology except for this one thing: the vast chasm is charted by the difference between faith and atheism.

140 http://www.britannica.com/topic/theology.
141 http://unabridged.merriam-webster.com/unabridged/polemical.

Act 3 — Scene IV

On the surface, theology seems to be anything but neutral. Remember its first roots: Plato made the point that theology as a discipline was **polemical**, in and by nature.[141] When the roots of a tree are oak, it cannot not yield apples. Consider. And take a moment to reflect and converse with your inner thoughts, do we not as a group of our own, tend to cling to what we know? Until, as James Burke so well noted, the "body of knowledge" changes and then we, too, can take on a new way of thinking. I am hoping that this book will do just that for many scholars to come.

One of the most basic ideas that I am intent upon getting across, is the idea of "relationship." When it becomes more obvious to the thinking mind that there is a vast amount of similarity in our "body of knowledge"—that all-inclusive, intellectual information—it also will become increasingly obvious that all that we experience and learn is not random. Instead, it is intricately designed and formulated to point toward that Power, that Force, that Being Who is the Creator of it all—and that is more exciting than one can know until one experiences it.

"In the beginning was the Word, and the Word was with God, and the Word was God. He was in the beginning with God. All things were created through Him, and without Him nothing was created that was created."[142]

142 St. John 1:1-3: *D. James Kennedy Topical Study Bible*: Charisma Media/Charisma House Book Group: 2015: Pg. 1502.

Scene V

THE ROAD LESS TRAVELED BY?

Consider this…

The science of Cymatics absolutely claims that without sound, all matter would cease to exist.

"In the beginning God created the heavens and the earth." Genesis 1:1
*Genesis is the book of beginnings. There is nothing else like it in all the world. It gives us the origins of **all things**.*
This verse alone—obliterates most of the erroneous philosophies that have come up during the history of mankind. Notice:

Atheism: *means "no God."*

Pantheism: *the belief that everything is God—if everything is God—everything is ultimately good.*

Polytheism: *there are many gods.*

Materialism: *the shared view of humanism, secularism and communism.*

Dualism: *two gods—one evil and one good.*

Evolutionism: *the belief that the heavens and the earth evolved: Genesis declares that God created them.*[143]

143   *The D. James Kennedy Topical Study Bible*: Charisma Media/Charisma House Book Group: 2015: Pg. 4.

To refute any of these statements is to challenge the veracity of
God Himself and His written Word.[144]

he critical conclusion to be drawn—and it is my hope that this book will be at least a piece in solving this puzzle, if not the final conclusive bit of reasoning in the whole matter—deals with the very truth of the existence of God Almighty. Without that one decisive conviction—there is a God and He is a Creator God—no author is able to make a strong case, or **any** case, for that matter, for the creationist approach to our universe and all that is in it. The humanistic and secularist logic, thereafter, is left quite intact. Our culture has done a very good job of eroding the foundations of our national faith and the faith that undergirds Judaic traditions—and in providing hundreds, if not thousands, of alternative arguments to support that erosion. Nevertheless, the position of this book is based on a Judaic/Christian platform. The Biblical account of creation is the core of my theory and, therefore, I must adhere to it closely, and without trepidation.

Theology is, in fact and by definition, "the study of God and his relation to man and the world: a branch of systematic theology dealing with the arguments for the existence of God, the divine nature and attributes."[145]

The body of knowledge does change, and the interpretation of that knowledge changes drastically, sometimes to the point that it actually becomes redefined. And so, the study of theology has seriously expanded over the centuries and has begun to include disciplines which would have been foreign to the early predecessors of the idea of the study of God. But, we dare not stray too far away from the original idea that theology is

144   My Note.

145   http://unabridged.merriam-webster.com/unabridged/theology.

the study of "God and His relation to man and the world." It is, therefore, my intent to keep to the simplest possible lines of expressing this phase of my theory which I believe has the potential of providing many answers to many questions and inquiries about origins.

> *"In the beginning God created the heavens and the earth."* Genesis 1:1

> **Genesis is the book of beginnings. There is nothing else like it in all the world. It gives us the origins of all things.**

There seems to be a great gulf fixed between the idea of things temporal and things eternal. Even as a young child, I would begin to think about the world and how huge it was, and then I would think about the skies and how much bigger they were than the earth where I lived. Inevitably, I would begin to talk with one of my imaginary playmates about this. This was long before, in my adult life, I began to allow Jana to shape the thoughts that were too fantastic for an educated, mature woman to formulate.

What I am suggesting is that I firmly believe in the power of the imaginary boundaries of the human mind which are **not** bound by convention or formal education. Einstein would not have become Einstein if he had not learned to collude with his mind of imagination. Jana gave me the privilege to think thoughts beyond the confines of the conventional, and for that I am forever grateful. It is true that the imagination must be kept in balance so that it does not begin to dominate our thinking. Then it can become out of balance, and it may even be difficult to tell the difference between those valued inspirational thoughts and pure egocentric fantasy, a quite unhealthy state of mind.

GENESIS 1

In the second verse of Genesis 1, we are told that the state of the world before creation was a state of formless void. We are not told what that meant, but is it really necessary for us to know what that void was like? What is necessary to understand, is that over that void, with a perfect plan already in place, the Spirit of God Almighty "hovered" or "fluttered."

One of the earliest lessons we learned about Biblical Hebrew at Liberty University was that Hebrew is a fluid, movable and changeable language. It is not static, in any sense. In a very real sense, it lives. In fact, the very meaning of words could change, and rather dramatically at that, simply because of the context. It was several years after my time at Liberty that I really began to understand the depth of that concept.

The Hebrew word for "hovered" or "fluttered" is *rachaph* (רחף). Its principle definition is "to relax, to flutter, to move, to shake." But, by inference, it can mean "to become loose." Have you ever had the incredible experience of having a front-row-center seat at a vocal concert given by, say, Paul Plishka or Beverly Sills? The trained singer knows that the only technique for successful execution of any intricate, long phrase is to fully relax the body while expelling all air from the lungs and banishing tension from the surrounding ribs and tissues. This allows the lungs, then, to completely fill with the air needed to float a beautiful, supported, focused sound. Could God have given us that stunning paradigm to help explain what happened when He uttered the first creative tones that ultimately became the shapes and matter that filled the void that existed before His glorious song was sung?

There is also the concurrent meaning of the word *rachaph*, which is "to nurture" or "to hover over," as a mother eagle would hover over her chicks. What an incredibly beautiful analogy, which maybe only a singer can fully grasp. As a singer I can attest to the laborious struggle that we all know as neophytes, how to do everything just right in order to make that one marvelously shaped and perfectly projected sound—and how glorious the satisfaction and joy when one knows that one has accomplished that sound—finally—What a struggle it is to be forced to work in order to relax.

With God Almighty, there was no struggle, no effort, no trial and error, because He was perfect God and perfect Singer God, Who needed no practice; the elements of the procedure were constant with Him. The problem is that so few people are willing to take the time to imagine and see in the spirit's eye the incredible beauty and the incredible parallels that are, really, right in front of us—seeing, we see not.

And so it went, for six days, that God sang and the universes and the galaxies and the wonders found in them were piled one on each until the work was done. An inconceivable opera of creation, with all of the facets and elements, sounds and rhythms perfectly in place!

Genesis 1:1

"In the beginning God created the heavens and the earth."

This verse alone—obliterates most of the erroneous philosophies that have come up during the history of mankind.
Notice:

Atheism: means "no God."

Pantheism: the belief that everything is God—if everything is God—everything is ultimately good.

Polytheism: there are many gods.

Materialism: the shared view of humanism, secularism and communism.

Dualism: two gods—one evil and one good.

Evolutionism: the belief that the heavens and the earth evolved: Genesis declares that God created them.

The only truth that must be dealt with at this juncture is that the first verse of Genesis puts to rest all other resistance to truth; one either takes it as truth—that God is the only God and that He created in the beginning or—one does not take it as truth and chooses thereby to believe that He did not create all that is created. Beyond that, there is no further truth available to us.

The great challenge is that if one does accept Yahweh as the Almighty Singing Creator God, then it only follows that one must trace the path of that creation through Scripture and begin to understand the ramifications and the wonder of that path. It is a creative path, of a certainty, and may never have been a path that we would have chosen to follow—the road not taken.

The astounding inevitability in all of this is to suddenly be able to see that all knowledge is bound up in this idea of the unity of God and His creative plan. That accounts for the almost terrifying similarities among all disciplines comprising the body of knowledge that we have come to in the 21st century. Every new discovery in science, for instance, causes a great stir among the constituents, and people at large are awed by the great intellects that make the discoveries. Not to belittle the great minds of our world; but stop to think about it—which of the great discoveries in history up to this day have been created by the awesome minds of the involved scholars? No, rather they amount to great findings of people who dared to allow themselves to **imagine** what it would be like **"if"**—

This is the point at which I am convinced that the next inevitability is to trace the development—the "evolution" (in the proper sense of the word) of that creative path through theological logic to see if we can apprehend the wonder of God's inerrant plan for the awesome world He designed for mankind.

The Road Not Taken

Two roads diverged in a yellow wood,

And sorry I could not travel both

And be one traveler, I stood

And looked down one as far as I could

To where it bent in the undergrowth;

I shall be telling this with a sign

Somewhere ages and ages hence:

Two roads diverged in a yellow wood, and I—

I took the one less traveled by, And that has

made all the difference.[146]

The genius of poetry is that it spans deep chasms in wisdom and knowledge. For me, in this present state of learning and apprehending, those two roads are: the first road which is bounded by human logic and the second road which is bounded by faith—the road of true theological understanding. This is the road which describes the relationship between God and His human creatures, whom He endowed with the most unique of all gifts—the gift of song. The living creature with the galactic ability to sing, also has, along with that gift, the gift

146 *The Poetry of Robert Frost*: Holt, Rinehart and Winston, Inc.: 1969: Pg.105.

to reason, and to comprehend so many of the complexities of this vast cosmos which he calls home. He can apprehend the profound mystery that music is everywhere, and, in fact, the human mind is beginning in this 21st century to see—or hear—or both—that indeed it is sound that holds that cosmos in balance.

Isn't it interesting that in ancient Hebrew rabbinical lore, it is well accepted that the very first of all creations was the creation of the Leshon Chadesh? Rabbi Mordechai Kraft teaches a very commanding lesson: the Biblical Hebrew language is the DNA of creation.[147] This concept has already been explored in more detail in the first Act of *God's Opera*. But it is of great importance, at this point, to reinforce the idea of language and sound as the DNA of creation, in order to underscore the veracity of the theological path that we are following. It is a necessity to be able to see that Yahweh, God the Creator, had every minute detail in perfect order at the very instant of that marvelous outburst of holy singing that became the vibrations, the sounds, the impervious raw materials of the creation saga. And I believe that as this present age of technological understanding progresses and unfolds, it will become widely, if not universally, accepted that this is a veritable fact—without sound, matter would cease to be.

Now, this has deep theological ramifications. But this is the path of faith, the road not taken by the vast majority of intellectually profound thinkers. It seems very odd, very ironic, to me that it is the world of science that is toying with and experimenting with this evasive idea of sound being the cohesive factor. By tradition, science is not strongly evangelical—but they have isolated a truth that I don't believe has yet come near its fruition.

147 https://24jewish.wordpress.com/tag/rabbi-mordechai-kraft/.

Act 3 — Scene V

We are taking the "road less traveled by," and therein lies the great difference. Only by seeing the flawless sequence of creation as it is decreed in the Book of Genesis can we begin to observe how infinitely well all the pieces fit together. Genesis 1:29 reads:

> *"Then God said, 'See, I have given you every plant yielding seed which is on the face of the earth and every tree which has fruit yielding seed. It shall be food for you. To every beast of the earth and to every bird of the air and to everything that creeps on the earth which has the breath of life in it, I have given every green plant for food.' And it was so."*

Two things emerge: 1) **This is not the place** to bring out all the preconceived ideas for confrontational argument against the Scriptural account of creation; and, 2) **This is the important place** for sanctified imaginative thought which can, if not fully accept, at least give possible acknowledgement to, the Biblical account which has, in spite of continual and concentrated attack, survived and has even inspired parallel accounts in every ancient culture. The presence of parallel folklore has been a growing intrigue in my quest for truth. Since relationship is part and parcel with my theory of the Singing Creator God planning, intending and accomplishing all things, it is a natural to see the unfolding of lore in all cultures which attempts to explain any events of importance with such similar folkloric stories. And this certainly does not diminish the power of the Biblical accounts. In fact, in my mind, the similarities only serve to underscore the eternal truth of the Biblical accounts. The questions which mankind has asked for eons do not change much. Where did we come from, and where do we go?

I TOOK THE "ROAD LESS TRAVELED BY," AND THAT HAS MADE ALL THE DIFFERENCE.

By this point in this book, it is my hope and trust that the path of the "less traveled road" has been well laid out. Now is the time to journey down that road and to try to determine where it leads and what it means.

For me, one of the criteria for finding reasonable answers was to be willing to allow my sanctified imagination to see through the misty morass of conflicting philosophies, warring theories, and hostile views of any disciplines which seemed to be contrary to my own understanding of knowledge. It occurred to me somewhere along the way that it might be much more exciting to see unity and harmony and flawless design than to be bound to prove any point of view that I might hold sacred to my way of thinking.

It was not difficult for me to have the Holy Bible as **my** starting point, since my entire orientation is evangelical and conservative in nature. But what if it weren't? What if I were an evolutionist, for instance, or an "old-earth-er," who is duty-bound to give space to the idea of an evolved world rather than a spontaneously executed and created world? As a matter of fact, in the forty-five year process of planning and finally writing this book, I **was** duty-bound by my own sense of justice to consider the vantage-point of the person who did not have my deeply ingrained faith system. I ultimately decided that the road I must take was the only one with which I could be honestly involved and comfortable, and would then be compelled to leave the eventual evaluation of the worth of my theory to those who read and study it further down the road.

It is reasonable, I believe, to make the assumption that a large majority of the world will not easily accept the idea that God Almighty, after duly designing and providing processes for His creation, simply uttered a sound, as stated in Genesis 1:3:

Act 3 — Scene V

"Then God said—and there was," "and it was so."

The Hebrew word for "to say" (said) is *amar*. Another of its primary definitions is "to utter," and that word has implications of singing. At least, we are within bounds to make the assumption that it was, indeed, sound of one genre or another that was the initial energy of creation. And what is even more exciting, and we have already discussed this idea, is that today science is spending its most concentrated focus on the function of sound, vibration, frequency, pitch in the apprehension of all the universe, its concurrent facets, and, indeed, all matter itself. This is the essential thread in the mantle of knowledge of all matter—sound.

It was not until I came upon the incredible amount of investigation into the properties and functions of sound in the total scheme of things that I began to see much more clearly. It was not unlike the old-fashioned slide show where the slides, warming in the light of the projector, snapped into focus. At almost one and the same time, Scriptures aligned themselves in my mind as being necessarily related to one another. The spiritual relationship of Jesus Christ as the Word of God, i.e., the Sound of God, the Song of God, to the inviolable truth of the Son of God as Co-Creator with God the Father all began to fit together sensibly. And the fairly new unexpressed thought of Jesus Christ as the original sound as the Word of God became feasible.

This is the "road not taken," to be able to allow the idea to mature that it was, indeed, the sound of the voice of God in the Person of the Word of God that made the first "utterance," which became the vibrations that have framed all that there is. Look:

- Genesis 1:1: *"In the beginning God created the heavens and the earth."*

- Genesis 1:3: *"And God said—and there was," "and it was so."*

- Psalm 33:6: *"By the word of the Lord the heavens were made, and all their host by the breath of His mouth."*

- Hebrews 11:3: *"By faith we understand that the universe was framed by the Word of God, so that things that are seen were not made out of things which are visible."*

- John 1:1-3: *"In the beginning was the Word, and the Word was with God and the Word was God. He was in the beginning with God. All things were created through Him and without Him nothing was created that was created."*

- Colossians 1:16: *"For by Him all things were created that are in the heavens and the earth, visible and invisible. All things were created by Him and for Him. And in Him, **all things hold together.**"*

How incredibly overwhelming to learn that science now, in the voice of the newly-emerging discipline of cymatics, is making the unequivocal claim that without sound, all matter would cease to exist. And what could that sound possibly be? The voice of God…

Act 4
MUSIC: THE BEAUTY OF HOLINESS

Scene I

MUSIC—THE GREAT ENIGMA

> *Consider this…*
>
> **All life has emanated from the deep well of energy and power, which belongs to God alone.**

*M*usic is, indeed, an enigma—a riddle, a puzzle, a mystery. There are far more questions about music than there are answers.

For Robert Shaw, music was his religion; when he conducted, it was his worship. But Shaw had a unique approach to the music that he so loved—it did not necessarily take the place of his God, but, having studied under his baton for fifteen summers in Princeton, NJ, I had the distinct impression that his underlying religious conviction was that God was implicitly expressed and revealed in the great choral works of both the modern age and antiquity. The standard that he held for us as his singers, therefore, was always reaching toward perfection. He was constantly in quest of what he called "truth in music," and when his "people" finally would give him nearly what he asked for, he could not contain his delight. The opposite was also true if we failed!

209

Oddly, when we would ask the maestro what he meant by "truth in music," he would chuckle, shuffle his feet and say, with a twinkle in his eyes, "When you find out, let me know, won't you?"

What is it about music that demands so much effort, clarity, love, devotion and discussion from some of the greatest thinkers that life has generated? I believe that I have been granted a very tiny glimpse into the possible world where that answer will at last be granted to us. Music is intimately intertwined with the same intimations and questions that we customarily reserve for seeking after the Almighty and Eternal God.

Now, of course, I understand that there is a very grave danger in making such a statement. Once, as I was trying to describe to a pastor that I felt that the Scripture verse which states that if we do not give open praise to Almighty God, the very hills, trees and rocks will cry out in His honor. The result of that conversation was that I was tagged as a pantheist and was openly promoting the idea that nature was God and God was nature. I was eventually asked not to worship in that church again. Of course, I meant no such thing.

So, by making the statement that there is some link between the attributes of God and the attributes of music, I must be, also, very, very careful to make it clear that, in the first place, I do not worship music. In the second place, I must be very clear that God as Redeemer and Creator is the center of my individual world and life, and it is He, only, Whom I worship. Thirdly, I must make it equally clear that this entire quest toward "truth in music" has been to establish the understanding that when I am talking about the development of this theory of the Singing God and creation as a musical event created, in fact, with raw materials of a musical nature, I am talking about a "something" that I have never heard and have no idea about its sound genre. God's music, the intent of His music, is an unknown at this time, but I believe that science is getting very close to uncovering the roots of it in their newer discoveries about sound, vibration and frequency.

The inquisitive mind is both a gift and a plague. The love of music afflicts many people. I say that because when that love is so deep and so all-encompassing, there are times that the afflicted soul cannot rest. And then we begin to wonder: Is music beauty or is it plague? Here is another huge window open for misunderstanding to enter through. There is as much difference of opinion about music as there are people to engage in the discussion. And, even on the surface of things, some music is just ugly—ugly to listen to, ugly to play or sing, jangling to the nerves of sensitive souls— and generally unpleasant. Where does that idea fit in this consideration of music as beautiful or even holy? That is for another time—in another writing—except to say that, when all is said and done, I believe there will be very clear delineations to help us to hear the real "truth in music."

For now, take a moment with your own personal imagination to consider a phrase which is repeated over and over in the books of poetry in Scripture. *"O worship the Lord in the beauty of Holiness."*[148] Worship is intrinsically related to singing and music, so it can be inferred fairly, I think, that the admonition to *"worship the Lord in the beauty of holiness"* has that same intrinsic call to discern that God has some very deep standards of beauty and holiness for music, and is even the possible source of what we eventually will discern music to be.

Then, of course, there are unanswered questions surrounding the art of music from the purely human point of view, that have been passed down from the times of Plato and Aristotle. Fashions in music come and go. Like other great art forms, music has expanded and filled up the world

148 I Chronicles 16:20, Psalm 29:2, Psalm 96:9.

with sounds that are really not fully understood. What is music? Is there a delineation between "good" music and "bad" music? Some earnest souls have been convinced that some music is even tainted with evil.

And hanging over this entire subject like a mist, is the question, "And where did music come from—what is its genesis?" I quote Curt Sachs often: "However far back we trace mankind, we fail to see the springing up of music. Even the most primitive tribes are musically beyond the first attempts."[149] I quote him because he captures in two sentences the thought that has been driving my musical quest since I was just a child of ten. It has been driving me because even as a child, I was convinced that there was far more to the story and history of music than anyone had yet discovered.

As I aged and pursued the life of wife, mother, pilot, teacher, organist—whatever—this haunting query always hung just beyond my conscious reach, until I finally was dropped somewhat unceremoniously into music school by my husband, who was a very wise "man of the cloth" and seventeen years older than I. I had had, by that time, several heavy losses in my life and my share of musical disappointments, and I did not wish to go to music school at all. I wanted to study psychology!

This was all providential, and it was precisely where God needed me to go, for He knew every moment that was ahead of me—even *God's Opera*. It was the people who were charged with my advanced education at Westminster Choir College who at last sowed the seeds of lasting love for my craft, which would supersede pride and lack of knowledge and direction and would give me a path to follow. And then and only then,

149 Curt Sachs: "however far back..." etc. **Curt Sachs**, (born June 29, 1881, Berlin, Germany—died Feb. 5, 1959, New York, N.Y., U.S.), eminent German musicologist, teacher, and authority on musical instruments.

could my Lord grace me with my life's real work, not to necessarily be a pianist or organist or singer, but to be seeker after "truth in music."

In the sixth chapter of his profound book, *Music, Modernity and God*,[150] Jeremy Begbie speaks eloquently to some of the deep questions of faith and the possible resolution of problematic issues of our faith in our world using music as a template or, perhaps more accurately, a metaphor. In terms of my own journey of discovery and quest for truth, this book has opened a vast new field of inquiry which I dare not even approach just yet, save to say this: one of the difficult considerations with which I have grappled is to find a path leading to the understanding of music in a new way—a way that will finally allow us to see that there is a holy beauty that emanated from God, the Singing Creator in His song and that that song is indeed expressed also as the Word of God, Jesus Christ. And it is the energetic frequency of His voice that is the bedrock material from which this universe and all that is in it, is created.

Begbie makes an interesting statement in his book that I intend to pursue at a later time, but it is pertinent to at least mention it here:

> "This book is an attempt to allow music the opacity of its own voice in the midst of theological renderings of the story of modernity."[151]

This relieves us of the burden of proof for anything we may feel constrained to think or write about the meaning of, or necessity for, music, if we can but discover the binding ties between God, the Singing Creator and the

---

150 *Music, Modernity and God*: Jeremy Begbie: Oxford University Press, Great Clarendon Street, Oxford, OX2 6DP, UK: 2013.

151 *Music, Modernity and God*: Jeremy Begbie: pg 7.

song that He sang, the music of the "Song of Creation." I realize the danger of sacrilegious wanderings and in no way am inferring such a thing. On the other hand, to see the real, "truthful" music as the Song of God, which is but the framework and raw materials of all that we can see and experience, and to begin to know how to evaluate these hidden truths spiritually would be a whole new vision of that truth and would give music "the opacity of its own voice." In addition, perhaps, a window would finally be opened to being able to grasp the idea presented by so many but left unresolved—the real meaning of "truth in music." There is so much left unsatisfied.

These queries are the genesis of my own journey. As part of my preparation for my life's work, God granted me experiences that would make the most successful, professional musician smile with kind envy.

SPOLETO, ITALY

As a forty-year-old untrained singer, I was auditioned into the famous Westminster Choir, and as one of the members of that elite group, privileged to sing for two summers at the Festival of Two Worlds in the ancient village of Spoleto, Italy. We sang evensong every evening on the steps of the Duomo, a cathedral built without mortar in the days of Barbarossa. And we gave several full concerts in the sanctuary which had an amazing sound reverb.

Charles Ives wrote a setting of Psalm 90 for double choir—eight voices. This work was a staple of our Duomo repertoire for two summers.

The incredible discord of Ives' music was enhanced by the startling acoustics in that sanctuary. One of the enduring dialogues among theologians and musicians alike is the correlation of words and music in any serious setting of Scriptural material. This particular Psalm setting is one of the finest examples of that vicarious relationship. Ives incredible grasp of dissonant harmony as an expressive agent only paints in the deepest and most vivid hues, the depth of God's anger when He is disobeyed.

Late one afternoon, we were rehearsing Psalm 90 in the sanctuary. If my memory serves me well, we sang it a capella, with only the organ playing a low "C" pedal-point throughout—and very quietly. When we reached the point at which Ives chose to express God's anger with a tight harmony of eight voices piled one-on-the-other over a span of more than two octaves, from lowest to highest, an extraordinary thing happened. Subtly, as if from the deepest heart of the cathedral, we began to feel a rumbling as our director was meticulously tuning each interval. The rumbling grew until it was barely audible, but it could not be denied and it clearly was not an earthquake. We all heard and felt it under our feet. The chill that came over us all was palpable. Dr. Flummerfelt cut the singing off, and said, "I think we should stop now. We have just heard the voice of God." We went home in stunned quiet.

That sound, mysterious as it was, had a power that none of us ever tried to analyze. I don't recall that we even tried to talk about it for the rest of the summer. Now, in the light of Dr. Begbie's incisive take on the ability of intricate music to, by virtue of its unique qualities, occupy sound-space and time with great diversity of pitch and timbre without destroying any of its uniqueness is creating an entire set of new curiosities about the characteristics of music and its genesis, that permits it to operate on such a plane. And, what does that mean, really?

A CLOSER LOOK AT HARMONY

One of the attributes of music that is particularly interesting to me is the vertical structure of harmony. What do I mean by this? Play a middle "C" on the piano. It has an identity which is readily apparent and irreduc-

ible. Middle "C" is middle "C." Now add to it the "E" which is one-third above it. You now have an obviously new identity, which we call a major third. The newly added note in no way dilutes or changes the identity of the middle "C." It adds a new dimension, but it also creates another irreducible identity. One can differentiate the two notes, one from the other, but if you take one away, the identity of the interval of the third is erased. Neither note impinges on the other. Now add the "G" one-third above the "E." You now have a three-note chord called a triad. That is, in and of itself, a remarkable metaphor. Those three notes, taken as a whole, have a unity and irrefutable character of three-in-one. It takes all three to create the whole and any one of the three, taken away, would destroy the construct.

The amazing thing is that it is only music which has the capacity for two or more of its basic elements (like a "C" or an "E" or a "G") to occupy the same space at the same time without changing the basic character of any one of the three. True, the actual result of such a phenomenon is that we also, at the same coinciding time create a new identity, but we have not lost the original identities either. And the resultant new construct is irrevocably more powerful, more beautiful, more expressive than any of the single elements taken one-on-one.

For me, this shed a new light on understanding the character of God. He is one God, the Trinity, made up of three entities, each of Whom is vital, alive and unchanging. But only as the Trinity is He fully recognizable in all of His glory, and without any one of the Three, the total identity is no longer identifiable. On the other hand, taken individually, each of the Three has its Own separate work and office in the Trinity, just as the "C" is still "C," the "E" is still "E," and so on. Visual arts cannot do this. Just as soon as one adds a new value of color to an existing color, the entire identity is changed—yellow superimposed on red becomes undeniably orange.

By apprehending this idea, do we find a means of a new understanding of the possibility of "truth in music" which, in all likelihood, is otherwise not a viable consideration?

A CLOSER LOOK AT RHYTHM

Beethoven's *Missa solemnis*, as I have often reiterated, was, for Robert Shaw, a source of great apprehension. Shaw was able to be extremely honest with and about himself, when it came to his relationship with his beloved craft. I am being intuitive and not analytical when I say that I believe that Shaw was not musically apprehensive, but I do believe that he himself intuited something much deeper in this particular work than the human mind could fully apprehend. This "truth in music" that he talked about so often was immensely obvious in this great choral work.

Beethoven's capacity to use rhythm as an expressive element was quite astounding. And a further extension of his genius was the ability to marry this rhythmic expressivity to the harmonic and structural construct in such a way that they each complimented and strengthened the other, and provided a musical and spiritual means of communicating truth which leaves mere words stunningly inadequate.

While Beethoven's genius was so evident in the writing of his music, Shaw's genius was just as profoundly evident, in my estimation, in being able to defer his own musical intent to the degree necessary to allow him to divine Beethoven's intent—or any other composer's intent for that matter. He had the uncanny ability to see and hear beyond the obvious to be able to grasp the deepest nuances of meaning, especially in the monumental choral works. His compelling force was the music itself, and he had the absolute intuition needed for that insightful approach. In the *Missa solemnis: Credo*, such insight was a transcendent force as we began

to apprehend the depths of the *Crucifixus Adagio Espressivo*. We were called upon to experience the "endless death" of the act of crucifixion, and the only means for expressing that, we were admonished, was in the realization of the music itself. Shaw told us that we had just eight seconds to tell the world of the "instant nails that punctured flesh," and to reveal the horror of the "hammer stroke." The ingenious metrics of Beethoven's writing made this possible, but we, the singers, had to stay "out of the way." The tempo was a slow "four" meter. The down beat in a four-beat measure is always the strong beat—or so we had always thought. Beethoven had marked that beat as *sforzando*, meaning suddenly and forcefully loud. But the problem was that that particular downbeat was tied to a thirty-second note from the previous measure. Impossible! The genius, Shaw told us, was that Beethoven knew that the hammer fell just before the sound and full anguish of the stroke was absorbed by the hand. The resulting "tah-AH! ta-AH!" resounding from the tympani a fraction of a second before the down-beat painted the picture as no words could.[152]

Another dramatic exemplar of this genius is found in the *Dona Nobis Pacem*. Leonard Bernstein said of this last breath of awesome holiness in the *Missa*, "This piece doesn't end," Shaw added to that remark, "…peace will only come from above."

The part writing is, as well, miraculous. The first incantation of *Dona Nobis Pacem* was described this way to us who sang this miracle with Shaw in 1975: "(Sing it) Like spring—like children. It is peace that responds—nature's renewal—but with desperation." Shaw conveyed to us that this was not peace that just is—it is peace that is demanded and warlike—schizophrenic—and in the music of the *Miserere* one can hear 'war.' The secreted, but real, condition of man on this earth?

152 Missa Solemnis, Opus 123: Beethoven: Breitkopf Edition Nr. 29: Printed in Germany: pg. 99.

The second, and more frantic war cry commences in the Presto section. The fractured rhythmic constructs of this section portray as no word could, the anxiety and distraction of mankind screaming for peace. The cries metamorphose into weary whimpers contrasted against impassioned shouts for "Peace!" The presence of war persists. The echoes of war persist. The weary ones offer one more prayer for peace. In the quiet of the last quiet *donna pacem*. there is yet one more hope—"peace." Shaw's words: "We are left with green grass and growing things—but we WON'T BE HERE!"

The last orchestral outburst declares: "Life ends with death—to this world."[153]

SUMMARY

These are but two places where we might be able to garner some deeper than usual insights into the power and efficacy of music as a communicative force. But what we are really pursuing is the deeper than usual understanding that in order to have such communicative power and, certainly the power to communicate truth, music must be far more than a man-made, man-centered craft which was given as a gift to enhance our quality of life.

It is my contention and firm conviction that all life has emanated from the deep well of energy and power, which belongs to God alone, and that He saw fit to allow mankind the privilege of seeing into His singing heart.

---

153 The remarks in this section are taken from my marked scores used at the Westminster Choir College summer workshops in 1975 and 1979, and are direct quotes by Shaw from the teaching sessions.

Scene II

THE ESSENCE OF SEARCHING FOR TRUTH IN MUSIC

> *Consider this…*
>
> **The world today dislikes the concept of absolute truth because they believe it constricts what they consider pleasurable and worthy of man's effort.**

There is nothing more difficult to take in hand, more perilous to conduct, or more uncertain in its success, than to take the lead in the introduction of a new order of things.
*Machiavelli, **The Prince** (1513).*

orty years is a long time—in the calendar of mankind, at least. And that is how long I had been struggling to formulate what I believe. In fact, for the first twenty years or so of my quest for truth in music, I was not even sure what it was that I was trying to formulate. A critique of the decline of music in general? The decline of church music? An intellectually acceptable view of music as a vast life force? The universal language? The elements of truth as expressed by music?

No. None of these ideas would allow me to shape them into a viable hypothesis. Music, by its very nature, was much too cosmic to permit

being put into any logical shape that could then be analyzed and explained. And that is precisely the moment at which my direction changed.

Much too cosmic! That was the key!

Ideas of all kinds had been roaming around in my head for years, and suddenly I understood what I was supposed to write about—creation—and music! Curt Sachs, the famed German musicologist, had made a deep impression on me in terms of his own research on ancient music and the actual beginnings of music. This quote echoed my sentiments exactly; unbidden, it lodged in my brain:

> "However far back we trace mankind, we fail to see the springing up of music. Even the most primitive tribes are musically beyond the first attempts."[154]

Of course! I puzzled over "…the most primitive tribes are musically beyond the first attempts." Gradually, I began to understand what I thought he must have meant: "…earliest man had an innate ability to make musical expressions, even before the very first attempts at musical organization were ever conceptualized." Certainly, early man had found ways of music making long before it was thought of to organize it into recognizable, meaningful symbols.

Somewhere in that frame of time, I came upon Genesis 4 which tells of the birth of Jabal, Jubal and Tubal-Cain. From the lineage of these three brothers came the primitive seeds of man's labors—the keepers of livestock, the father of those who play on the harp and flute, and those who work in bronze and iron. Jubal played upon the harp and flute.

154 Curt Sachs (born June 29, 1881, Berlin, Germany—died Feb. 5, 1959, New York, N.Y., U.S.) eminent German musicologist, teacher, and authority on musical instruments.

Act 4 — Scene II

Jana and I spent many hours pondering the beginnings of this music; there was pitifully little detail in that 4th chapter of Genesis—just enough information to spur thoughtful imaginings about how Yahweh must have instructed these men—and especially Jubal—in their respective arts. There was no one else to teach them. Such heady accomplishments as building musical instruments and then learning to play on them well enough to teach others was no small consideration. And then—where did the musical sounds originate and how did they fall into such unquestionably orderly form as they must have done?

In the chapter on theology, we have already reflected upon the hypothesis that when God spoke—sang—the very first creation commands, He had already perfectly designed and organized all systems that eventually were in place for mankind to discover. Creation was, I believe, a complete and altogether organized and unified entity of intended perfection. So, Jubal had to work very hard, I imagine, to apprehend the qualities and characteristics of this instrument he had been told to construct. And then, he had to become proficient enough to make beautiful sounds on it and to teach others to play as well.

Ah! Hah! The virgin expression of musical beauty and order, and the birth of a consistent system of learning to make such beautiful music. No wonder that Curt Sachs could find no logical "springing up of music." That springing up of the wonder of music reaches all the way back to the days of Genesis.

There is a well-accepted perception in academic circles that music is a human activity, designed and brought to fruition by human effort. It is also accepted that, since man is the only creature capable of such intellectually superior efforts, he must retain a very elevated position in the spectrum of creative genius. In fact, he, mankind at large, may have begun

223

to think of himself as co-creator, and such thinking made it possible, by the Age of Reason—the "enlightenment"—to see himself as the center of creating and learning. This is where the shift of paradigm began to effect the ways in which we now see the worlds of the intellect.

There was another brave attempt to explain the unexplainable that reverberated up and down the halls of musical acuity and astuteness: music and its value were tentatively made clear to us. Although of questionable value in real intellectual terms, it was, indeed, of some philosophical value and could even be viewed as more or less necessary in order to enhance the quality of life for those who listened or, better yet, participated. It improved the aesthetic quality of life. Of the pages and pages of information I have read on this subject, I have never discovered for sure what that meant. Neither have I been able to ascertain what the musical authorities finally understood music to be and how it accomplished what they thought it accomplished—or maybe even to define **what it actually did accomplish.** The unrelenting **"Why?"** of music was never revealed satisfactorily to me.

One highly respected music teacher, philosopher and writer, Leonard B. Meyer wrote:

> "A good piece of music must have consistency of style: that is, it must employ a unified system of expectations and probabilities; it should possess clarity of basic intent: it should have variety, unity, and all the other categories which are so easy to find after the fact. But these are, I think, only necessary causes. And while they may enable us to distinguish a good or satisfactory piece from a downright bad one, they will not help us very much

when we try to discriminate between a pretty good work and a very good one, let alone distinguish the characteristics of greatness."[155]

Roger Sessions made a similarly vague assessment of the musical experience:

"No one denies that music arouses emotions, nor do most people deny that the values of music are both qualitatively and quantitatively connected with the emotions it arouses. Yet it is not easy to say just what this connection is."[156]

All very interesting and probably very true—but my gnawing and inexorable question remained unaddressed: **After all, what is music and why is it the most persistent, insistent and ubiquitous of all the arts?** And where did it **really** originate? Or, did it just evolve, like everything else just **evolved**? And, **how** did the musical rules and guidelines develop; who invented the triad; who coined sonata form? Oh, yes, it is very easy to say that these things just developed over time as the need arose because they were so logical or so functional. Without a firm foundation of truth and logic, none of these "answers" really "answer" the questions that surround any discussion of music at large. In addition, the subject is itself so vast that without understanding what the roots of this tree are, naming the fruit is an impossibility.

---

155 Leonard B. Meyer: *Music: The Arts and Ideas*: Chicago University Press. 1967.

156 Sessions, Roger: *The Musical Experience of Composer, Performer, Listener*: Princeton, New Jersey: Princeton University Press. 1950, republished 1958.

Graduate school at Montclair State University in New Jersey was a sifting time for me. One professor in particular stood out in my mind. Dr. Rosalie Pratt constantly challenged me to clarify my thinking and not to allow myself to be generic or stereotypical in my thinking. Even when she somewhat fiercely disagreed with some of my ideas, she encouraged me to think, think, think, and for that I thank her posthumously. It was during this time that I began to formulate what would become the cornerstone of my thesis, though at the time I did not realize it.

I only vaguely recall the actual classes that I had under Dr. Pratt, but the content of our relationship was the lynchpin. I remember disagreeing strongly with the philosophical premises of Leonard Meyer's writings on the uses and worth of music education. They seemed so empty that I had to sympathize with the students who sat in "music appreciation" class and threw spit-balls. None of the cotton-candy explanations of how music enhanced our human condition and gave reason to our "artistic" lives carried any weight for me, nor did I believe that they had any real logic behind them. That was in the years from 1976-1981. Little did I realize then that it would be 35 years before I could link that idea to the idea that the Age of Reason would eventually enlighten us to the degree that we humans could begin to see ourselves as the creators. That realization would open my mind to a growing glimpse of truth.

A PRESENT-DAY EPIPHANY—MUSIC IS NOT, AT ITS HEART, A HUMAN ACTIVITY

Be careful now. It is dreadfully important not to misplace the importance of human accomplishment. As has been noted earlier in this book, the Age of Reason was essential to the survival of mankind. Good of many kinds, progress, advancement and even improved human dignity have resulted from the philosophies of the Enlightenment. But the shift of paradigm that accompanied that radical change of philosophical momentum has blinded mankind to his finiteness. We have largely forsaken the idea that we are, after all, finite, mortal and, therefore, limited beings and not demigods. To be able to explain in purely human vernacular something as ubiquitous and over-poweringly present in every place and situation as music is not only presumptuous, but also impossible. Even as a student, I viewed Meyer's attempts at explaining the elements that described "good" music as being far short of what my intellect needed to satisfy it. It was frustrating to me.

And here is the pivotal point. With man as the centrality instead of God, everything takes on a mortal mantel. In ACT III, we looked at Jesus Christ as the Word of God—the Song of God—the original and formative **sound.** In ACT II, we explored the concepts of sound that have greatly increased our understanding **of** sound—vibrations, frequencies—as the cohesive forces in all of nature and the universe. The Word of God—or the Song of God, if you please—infers **that** concept of sound. And Scripture itself, as we saw in ACT III, is very clear that in the Word of God—Jesus—*"all things consist."*[157]

On the other hand, if we take that element out of our narrative, then everything necessarily reverts back to man as the centrality, and, therefore, the primary creator and inventor of music—and other things, as well. In such a case, nothing can be explained beyond the **mortal.** And it is quite basically clear to me that without the element of eternity and omniscience, mortality undercuts all other rationales.

---

157 Colossians 1:16, 17: NKJV.

THERE IS NOTHING MORE DIFFICULT TO TAKE IN HAND, MORE PERILOUS TO CONDUCT, OR MORE UNCERTAIN IN ITS SUCCESS, THAN TO TAKE THE LEAD IN THE INTRODUCTION OF A NEW ORDER OF THINGS.

And yet I am compelled to take this stance and to make this statement: There is, outside of seeing God Almighty as the ultimate and final Creator and Jesus Christ as Co-Creator with Him from before the beginnings of time, absolutely no explanation for the intricacies and complexities of creation. And, even more than that, to consider God Almighty as the Omniscient Singing God, Who used the elements of music as the raw materials of creation is equally necessary, based on what we can learn from the Hebrew language, the science of sound and vibration, the Holy Writings and from music itself.

As the author of this writing, I am well aware of the perils of making such a sweeping statement, but it is my hope that as the reader considers the fabric of music as it unfolds in the next few chapters, the very beauty of the idea of music as the raw materials of creation will unfold with it and will make its own case for truth in music.

Today's world dislikes the concept of absolute truth because it constricts almost everything that today's world considers pleasurable and worthy of man's efforts. But the concept of ultimate truth is a concept centered in the whole perception of God as omniscient, omnipresent and eternal.

Consider: there is music everywhere. If it is, indeed, what I believe it is—the analogy given to us by this perfect Singing God as the raw materials of His creation—that idea in and of itself, would explain the all-permeating nature of this thing we call music.

Scene III

TRUTH IN MUSIC IN NATURE?

Consider this...

No matter how far back you look, we cannot trace the beginning of music because it was the raw material of creation.

"There is, indeed, little of what can be accurately called music in nature, for music is the Divine prerogative of human and angelic beings, and nature furnishes only the rude elements of music, the uncut diamonds, as it were, of sound." Frederic Farrar[158]

"The silence of the world. What do we hear when we hear music?—is music just a sound among other sounds? Do we consider music a mere continuation of worldly sounds? The song of the bird can be recognized as a first promise of melody, but the progression to music is seldom reached by imitation." Ruud Welten[159]

O, worship the Lord in the beauty of holiness, tremble before Him, all the earth. Psalm 96:9

---

158 Frederic William Farrar (Bombay, 7 August 1831—Canterbury, 22 March 1903) was a cleric of the Church of England (Anglican), schoolteacher and author. He was a pallbearer at the funeral of Charles Darwin in 1882. He was a member of the Cambridge Apostles secret society.

159 http://www.academia.edu/1269970/What_do_we_hear_when_we_hear_music_A_radical_phenomenology_of_music.

229

God's Opera

*T*he statement above from Frederic Farrar, Archdeacon of Westminster Abbey in the late 1800's, is the complete antithesis of all that I believe, and it was made by a man who was a pallbearer at Darwin's funeral. If God Almighty did, in fact, use vibrating sound, the sound of His own voice in the Person of the Word of God, Jesus Christ, to create all things from nothing, then it follows that the most primitive of those sounds—like the sound of a nightingale—would, logically, be musical and beautiful, and holy in nature. And we can infer that all of the sounds of nature—birds, rain, the wind, the babble of a brook—would carry the thumb-print of that Creator, and that would mandate beauty and harmony—and music!

Centuries prior to Farrar's birth, St. Francis is said to have made a lovely observation to one of his monastery brothers. They were sitting together on a bench in the outer garden, and a nightingale flew by, singing as she went. Then she returned, and St. Francis commented, "Let us sing praise to God antiphonally [alternating] with that lovely bird." He heard music that, obviously, the archdeacon did not. It is my belief that **that** beauty and **that** music are mutually **inclusive!**

How, in the face of the overwhelming beauty that St. Francis could see without even looking, beauty in every bird-song, in every whisper of the willow tree singing in the soft wind of Assisi, in every sunrise over the easterling Umbrian hills, could a man be so obtuse and so insensitive as to say, "There is, indeed, little of what can be accurately called music in nature"—or—"Nature furnishes only the rude elements of music"? The indication is that the deeper sense of truth in music is missing in so much of our intellectual exchange **about** music, and so it has been for all the ages.

Exploring this dichotomy and finding a living, singing refutation lays the foundation stones for a structure that I hope will become a beautiful monument to the truth and beauty of the music of nature. For it is my observation, along with hundreds of other seekers after truth, that, indeed, the incredibly exquisite sounds of nature are only a precursor to the other musics that we shall explore. But nevertheless, they **are music.**

The vast magnificence of a Bach chorale sung by the pure unspoiled voices of a boys' choir contradicts the possibility that the elements of the music of that chorale could have been invented by mere man. To listen to that music is one thing: to perceive how the elements that make up that music came into being in the **first place—the beginning**—is quite another. And I am not referring to the **composition** of the piece. I am referring to the actual embodiment of the musical work in terms of the sounds, vibrations, frequencies and pitches that make the music and how they were originally crafted. That is beyond human understanding.

How did the body of the singer become equipped to change pitch so perfectly and in addition, how did the particular, peculiar change in frequency come to create such perfection of sound and harmony in the first place? What are the mechanics of the breath and its physical components? And how does the action-reaction of the human breathing apparatus fall so flawlessly into place in direct parallel to the music itself? There are too many rabbit-trails to follow to be contained in a book of this kind, but, remember, it is only my intent to inspire and prod curiosity, not to solve the dilemma.

> The silence of the world. What do we hear when we hear music?—is music just a sound among other sounds? Do we consider music a mere continuation of worldly sounds? The song of the bird can be recognized as a first promise of melody, but the progression to music is seldom reached by imitation.

It is astounding to read some of the ruminations of self-appointed scholars and philosophical thinkers from the 20th and 21st centuries. I say "self-appointed" because they openly admit, at least some of them, that their ideas are exclusively subjective and even identify their work as **being subjective also**.

This preoccupation with the accomplishments of mankind and his overwhelming ability to invent and produce, I believe, is a direct result of the impact of the Age of Enlightenment. The paradigm shift that took place is profound, slow to develop and by reason of necessity, is not often questioned. If we were to take an objective look at the changes in philosophy that followed the dawn of the Age of Reason—the Enlightenment— we would find, I believe, that it is a logic of the human mind to begin to elevate the place and posture of mankind to a higher stature and maybe even a greater stature than God Himself. Which is not to say that man's stature should not have been improved, but to begin thinking of himself as more lofty and accomplished than he ought to think is quite another proposition and would require another book of inquiry in order to deal with it.

What is distressing, however, is that the natural outcome is such a change in perception that one of the foolish results is the reversal of man's acceptance of God as All-in-All: Creator, Teacher, Savior, the Great Designer and Maintainer of the universe.

Music was already being "performed" by God's creatures before the advent of Jubal and his commission to create musical instruments and then learn to play them and to teach. That is where we must start.

O, worship the Lord in the beauty of holiness, tremble
before Him, all the earth.
Psalm 96:9

Act 4 — Scene III

Why would the earth tremble before the God of creation? That really is a rhetorical question. I recall one time hearing Robert Shaw say, hanging his head almost as if in shame, "Every time I am faced with conducting—or teaching—the *Missa solemnis*, I shake in my boots. I am terrified!"

This great man understood truth and beauty in a way that few of us are gifted to understand it. It was not the greatness of Beethoven nor even the music itself that so intimidated Shaw. It was the abject beauty of the truth for which it stood, and he well knew the total human inability to grasp that beauty and truth, let alone to express it adequately. There was nothing superficial about his humility. And that is precisely what gave Shaw the incredible power that he had to allow that very power to show through the music that he so passionately loved; it was, indeed, his way of worshipping to be able to step into that evasive world of unspeakable beauty and share it with us and the listeners.

BIRDS

If Robert Shaw trembled before his corner of the creative beauty of the world, why wouldn't the world itself tremble before that utter, inexpressible beauty of the creatures that God Almighty gave her to shelter? The birds for instance. And how dare anyone say that the sounds of birdsong are not music? All one has to do is to listen. There are birds found all around us—everywhere! And it is a narrow sense of musical structure and beauty that permits the idea that if it does not fit the template of what **we know** as music, it is not music.

Ounce for ounce, the energetic force of a cardinal marking off his territory could cause a heldentenor to quake. The entire little red body is very obviously engaged in making that glorious sound. Or take the wood-thrush. His voice can sing two notes at once. Not even the greatest

coloratura soprano can do that, nor can she sing up the scale and down at the same time. Take note, too, of the infinite variety in birdsong. The blue bunting is capable of regional dialects, just as humans beings are. From mountain to mountain, its songs can vary.

As a matter of fact, some of the world's greatest composers have often found their inspirations in the songs of birds. Resphigi penned *The Birds* and *The Pines of Rome*; Ralph Vaughn Williams, *The Lark Ascending*. But the most captivating of all is probably the 20th century French composer, Olivier Messiaen. The following is an excerpt from Cornell University's awe-inspiring website, AllAboutBirds:[160]

> Most birders carry a notebook with them to jot down observations or to quickly sketch a bird they see—but one French bird watcher sketched the songs themselves, using his incredible ear to transcribe nature's notes onto a musical staff. That man was Olivier Messiaen (1908–1992), one of the great composers of the twentieth century.
>
> Messiaen's unique talent emerged early. As a boy, he honed his ear listening to and transcribing the songs of birds at his aunt's farm in the French countryside. He first used birdsong in his work during the darkest period of his life: while imprisoned in a German prisoner-of-war camp during World War II, Messiaen composed and then conducted, with a prisoners' quartet, Quatuor pour la Fin du Temps (Quartet for the End of Time, 1941). One of his most celebrated and important compositions, the somber

160   https://www.allaboutbirds.org/concerto-for-wood-thrush-and-oriole-bird-songs-in-classical-music/.

piece includes stylized songs from the Eurasian Black-bird and the Common Nightingale. The third movement, Messiaen later remarked, represents the "abyss of time, its sadness and weariness. But contrasting this theme are **the birds, who are the opposite of time; they are our desire for light, for the stars, and for all things sublime**."

After Messiaen was married, he would ask his wife to go out into the garden, listen to a particular bird song or call, write it down and then play it for him on the piano so that he could replicate it in a composition. It might have seemed like an odd fixation, but the beauty of composition that resulted is testimony to a talent so rare that it must have been God-given and God-inspired. It is one of my deep beliefs that the Creator God does, in fact, grant to certain of His chosen vessels rare abilities like this for the purpose of underlining His creative ends, and granting to the rest of us a gift just as rare to be able to partake of that beauty.

In direct contrast, there is the philosophy of Ruud Welten who claims that a bird's song can provide us the first promise of melody or music, but that the progression from there to music cannot be accomplished by imitation. And I believe that he is intimating that bird song is only imitative. According to the research done by Cornell University and public radio's *Greene Space* in New York City, that is not factual.

Dr. Sarah Woolley, Columbia University Psychology Department, New York City, has completed a fascinating series of experiments and observations on how songbirds develop and use their singing capabilities. Her lectures are broadcast on public radio's *Greene Space*.[161]

161   https://www.youtube.com/watch?v=uMf4OShCCMY.

The newly-hatched birdling begins immediately to memorize his father's song. The first 20 or so days of the young bird's life is critical to the learning process. After that, the father can be taken away from the nest and the offspring will continue to practice the song until he reaches sexual maturity and then he will begin to improvise upon the song until he has perfected his own version. This is his own individual improvisation on his father's personal song. Each species has its own characteristic song which is never violated, but three brothers from the same nest at the same time will have three differing versions of that song. Imagine! A yellow warbler's song, is a yellow warbler's song, is a yellow warbler's song and yet, three brothers will have three versions of that song. The immense provisions for variety in this sense are mind-boggling to consider!

To absorb this scope of singing sophistication in a tiny bird which weighs, maybe five or six ounces, contradicts the idea that the songbird's song cannot be music. Furthermore, Dr. Woolley makes this observation:

> "Songbirds learn to recognize, respond to and produce the complex songs of other birds. They use their songs to communicate socially, in the contexts of mating and self advertisement. This makes them very interesting model systems for understanding how sensory signals are encoded and decoded by the brain and how that process results in perception and social communication."[162]

These studies also make it clear that socialization is not the only use for songbird's singing. They communicate alarm, territorial boundaries and mating protocols. These particular songs are understandable between

---

162   http://www.columbia.edu/cu/psychology/fac-bios/WoolleyS/faculty.html.

different species, and the birds do exchange singing to convey safety signals, but reserve all other singing for their own. It is interesting that the female does not sing, but she has had to learn just as much as her male siblings because she has to recognize the various calls of her species as apart from all other species.

Taking into consideration the myriads of species and the thousands of members of each one, it is almost essential to believe that birds have a very well-developed kind of cognitive ability, and that ability infers the ability to make music. That being the case, it causes me to smile to understand that man is unwilling to admit that a mere bird could be gifted with song as well as we are.

This is but the tip of the iceberg of wonderful knowledge of birds and their music. They are meant for our enjoyment and amusement, but there is so much more to the life of the "'common" bird. They live in community and accept responsibility for one another in that community; and their communication is musical in nature—no pun intended.

WOLVES AND DOLPHINS

Nottingham Trent University in the UK has been conducting a study of the cries of wolves. Researcher Holly Root-Gutteridge wrote this for the *Daily Mail*.[163]

> "'It's (the wolf cry) a bit like language: if you put the stress in different places, you form a different sound.'
>
> The study, published in the journal, Bioacoustics, recorded eastern grey wolves whose howls can travel up to five miles and are used to defend territory from rivals and to keep in contact with other members of the pack."

163   http://www.dailymail.co.uk/sciencetech/article-2373903/Wolves-unique-singing-voice-identify-howl.html.

We can infer from this quote that, like the yellow-throated songbird, wolves have a highly developed system of communication and the quality of that communication is musical in character.

And the uses of those cries or songs are complex. Each wolf has its own particular **voice**, which is to say that even we humans can distinguish between one animal and another, and, of course, the wolves themselves, then, must have a sophisticated appreciation for the sounds of the voices of their pack mates. This gives them a fail-safe method of communicating and understanding the needs of the community.

We human beings have developed, over the years, certain preconceptions of the things around us, and, perhaps, wolves are not among the creatures that we prefer to associate with beauty. Actually, they are graceful and beautiful creatures, and aside from their natural predilection for the farmer's small animals, they are highly interesting and not to be necessarily disliked. The movie, *Dr. Zhivago*, gave me a new appreciation for the animal. I could well understand how Uri, hearing them sing in the white moonlight of snowbound Siberia was deeply moved to poetic utterance by their song.

Root-Gutteridge actually did her doctoral thesis on the songs of wolves and spent literally hundreds of hours studying and analyzing thousands of samples of their singing. Her essay "The Songs of Wolves"[164] is laden with intriguing and captivating information that makes it quite clear that the language of wolves, as well as being very musical is intricate and territorial in addition: "Like the difference between Farsi and French," she writes. Interesting that she sees their communicatory hallmark as being "musical."

---

164   https://aeon.co/essays/we-learn-more-about-our-language-by-listening-to-the-wolves.

Act 4 — Scene III

She makes the comment, "Songbirds display particularly complex rules to the order of their singing notes," making it quite clear that she recognizes on some level that the musical component in animal communication is a reality, and that, furthermore, this knowledge might, and probably will, give us important insight into human language, as well.

In the article printed in the *Daily Mail*, Root-Gutteridge also writes about the vocal prowess of dolphins:

> "—dolphins have names for each other, and can call each other just like humans, say scientists. Instead of words like 'Alan' or 'Bert' they have specific signature whistles for loved ones and social companions — the only animal species apart from ourselves known to do this.
>
> "A study of wild bottlenose dolphins off the east coast of Scotland found they responded to their 'own' whistles by calling back.
>
> "The findings suggest dolphins use signature whistles as labels to address or contact individuals of the same species they meet at sea."

The findings of science around the world underwrites the belief that the animal world and others exhibit a profound awareness of and capability to communicate and to enjoy, and that for the most part, those communiqués are musical in nature.

It is a matter of constant interest to me that scientists consistently use musical metaphors and other examples when describing so many of their experiments and findings. It has long been a premise of mine that

music—the music that we understand—is the only art-form that carries the most ubiquitous reputation among all other arts. One does not generally find dance in the practice of medicine—but one will often find some musical form or other in that practice. Poetry, as beautiful and beneficial as it is, is not often put to use in the designing of a bridge or building. But, if one looks carefully, one can find musical elements such as rhythm in the shaping of dimensions, and, most certainly, the existence of harmonics as a vital component in the successful accomplishment of any engineering project. Even the string theorists use musical descriptions and metaphors in making their difficult equations and theories more understandable to the vast majority—even to themselves.

We could keep going in the attempt to fix the idea of music in nature, and especially in the animal world, but actually I am only attempting to initiate interest in more discovery of the existence of truth in music—starting with the natural environment.

> *"For you shall go out in joy and be led forth in peace; the mountains and the hills before you shall break forth into singing, and all the trees of the field shall clap their hands."* Isaiah 55:12

> *"Oh, sing to the Lord a new song; sing to the Lord, all the earth!"* Psalm 96:1

> *"Let everything that has breath praise the Lord! Praise the Lord!"* Psalm 150:6

As a young woman, I would read these words, and, as a musician, I would think, "How quaint!" to think of the mountains and hills singing to the glory of God! It was a lovely picture that made me feel at peace in my soul—I could imagine that all of creation was and would forever be involved in the praise of God. But it was also something like a fairytale illustration. Very beautiful but also very detached from what I knew as the real world. The research for this book has proven otherwise—the idea of all of nature resounding with song in praise of the Almighty Creator God has become, not only feasible, but also incredibly beautiful as a real image of the music found in everything that has life.

I could spend the next twenty pages retelling what artists and philosophers and scientists and mathematicians have already discovered and have already documented. I could quote the poems of great poets like Robert Frost and it would all be good.

Lang Elliot,[165] naturalist, philosopher and writer, has made some of the most beautiful and grandest videos and recordings of "natural music" that I have ever heard and seen. It is really too bad that website links cannot be activated from the printed page, but below is the link that every reader should avail himself or herself of. Without the efforts of people like Elliot, some of us may never have the opportunity to drink in the beauty of the world around us.

There is a picture that I have carried with me for over 72 years. I was probably 12 years of age and finally had the opportunity to go to Girl Scout summer camp at the Bear Mountain State Park in New York's Hudson Valley.

The hills around Lake Sebago were heavily wooded and rich with flora and fauna of all kinds. There were Indian Pipes, a delicate and incredibly stunning fungal flower which looked like a tiny bell perched atop a thin stem. It was about four inches tall, and it grew on wet, rotting leaves shed

---

165 http://musicofnature.com/.

from last summer's growth on the huge, aged oak trees. The platforms for our tents snuggled deep in those oak and maple woods, and the growth was so thick that you could not see the tops of the trees.

Then there were raspberry and blackberry vines, heavy with fruit, and further into the forest a swampy area where the blueberries grew. And wild bachelor's buttons and purple thistle and wild lilies! It seemed that the flowers were all in a huge competition for Who Is Most Beautiful?

And the birds! How many different birds were there? Who knew? They sang all day long, and some of them sang all night, too!

One night, there was a thunder storm. God's tympani, rolling and crashing through most of the night—now right over head—now in the far distance. And the rain—did it really sing, too, or did my mind invent the ethereal song? What did it matter—it was God's symphony, orchestrated by Him and for Him, and for our excitement and pleasure.

That two weeks has never faded in my memory! It was a constant feast on the music of God—and since then, no one can tell me that nature does not sing. And, of course, God knew this when He inspired the writers to tell about the songs of the mountains, and the brooks, and the rocks, and the trees, and the wind—all singing praise to Him! Now, I could understand!

YOU ARE HELPING ME UNDERSTAND THAT BEFORE MAN HAD ANYTHING TO DO WITH CREATING MUSIC, GOD'S MUSIC WAS EVERYWHERE AND IN EVERYTHING THAT HE HAD CREATED. IT MUST SURELY BE IN EVERYONE ALSO, BUT NOT EVERYONE ALLOWS IT TO BE MANIFESTED THROUGH THEM.[166]

---

166 http://musicofnature.com/.

Act 4 — Scene III

It is a constant source of astonishment to me that the "common man" is so often the very one that God gives the purest insights to. Maybe that is because their minds are not so cluttered with self-acquired knowledge that any thought which departs from "the book" is anathema to them— as it can be to the great unwashed intellectuals whose greatest pleasure comes from what they **"know."** This is not to decry the wonder of "knowing," nor to decry those who do "know," but it can be a tempting notion to feel well-informed—better informed than someone else.

Therein lies the very inspiration for being in constant contact with one's innermost self who may be irreplaceable in helping one to keep his or her equilibrium concerning the things of eternal weight—such as the awareness of God's beauty and new ideas. Jana has often kept me from error.

Indeed, I believe that all nature was singing long before man came along and found a way to preserve the sounds of that singing by learning how to write the symbols of those sounds on papyrus to be remembered forever on this earth. But in our busy lifestyles, we have forgotten how to **listen** in that primitive way. Another thing that perhaps we have forgotten is the meaning and use of words. To the 21st century man, "primitive" may have a slightly negative aroma. Not necessarily so—in fact, the *Unabridged Dictionary* has as its first entry this definition for "primitive:"

> "not derived from or reducible to something else: ORIGI-
> NAL, PRIMARY <seeks excellence at its primitive source—
> nature — John Dewey> <an acre of one primitive color
> alone — J. A. Michener>"[167]

---

167 http://unabridged.merriam-webster.com/unabridged/primitive.

243

Music **is**, I believe, a primitive art. That is to say that it is "not derived from or reducible to something else."

Curt Sachs was absolutely right, I think. No matter how far back we look, we fail to trace the springing up of music—because it always has been. So, music is the raw material of creation, isn't it? Consider.

Scene IV

TRUTH IN MUSIC IN THE COSMOS?

> Consider this…
>
> *Music is frequency, vibration, pitch and sound, and it is the secret, unifying, integrating force that co-exists in every area of our universe.*

"Where were you when I laid the foundations of the earth? Tell me, if you have understanding. Who determined its measurements? Surely, you know! Or who stretched the line upon it? To what were its foundations fastened? Or who laid its cornerstone? When the morning stars sang together, and all the sons of God shouted for joy?" Job 38:4-7

"In string theory, all particles are vibrations on a tiny rubber band; physics is the harmonies on the string; chemistry is the melodies we play on vibrating strings; the universe is a symphony of strings, and the 'Mind of God' is cosmic music resonating in 11-dimensional hyperspace." Michio Kaku

In my sanctified imagination, where I often confer with Jana, I can easily see the question of truth in music first being posed in a most exciting and provocative way, as God the Creator and His

245

favored creature, Job, went head to head about the created world and its beginnings. The first time I read this story, I admit that I read it as most people read most Scripture, in a remote and maybe even detached way, as if it has no reality-quotient in our human venues. There is a barren and infertile attitude toward Scripture that keeps the energy and abounding inner life of the Word of God under a rein in so many ways.

I just read Job 38:4-7 to an unchurched friend of mine as it might have transpired between God the Creator and a contemporary string theorist whom I named and she knew about. The look on her face was most remarkable. I think she had probably never ever thought of God as conversing with anyone, let alone a well-known scientist who might be asking the kinds of questions that Job asked and certainly that all thinking scientists are still asking in today's world. And here is the quest laid out in simple terms: *"Where were you when I laid the foundations of the earth?—or, when the morning stars sang together?"*

You, see, in the very beginning, God the Creator knew all the answers to all the questions, and—He knew just exactly how to ask those questions in order to elicit the most incisive replies. Interestingly, there was often no answer fitting—not because there was no answer, but because the answer, when given to the Creator Himself, suddenly became all but irrelevant, because one found oneself standing in the presence of the awesome power that hung all things in space to begin with. Imagine!

I have watched and listened with interest as today's scholars have grappled with the issues of creation, matter and the universe—or universes, as their predisposition may transport them. Almost without exception, the greatest of those scientists and scholars have brushed shoulders with what I see as the absolute truth of Scriptural, theological reasoning as it concerns the origins of all things.

Act 4 — Scene IV

One of Stephen Hawking's most probing questions was:

> "Even if there is only one possible unified theory, it is just a set of rules and equations. What is it that breathes fire into the equations and makes a universe for them to describe? The usual approach of science of constructing a mathematical model cannot answer the questions of why there should be a universe for the model to describe. Why does the universe go to all the bother of existing?"[168]

In his personal journey of discovery, Hawking has trod many paths of belief, unbelief and honest questioning. But if one were to seriously consider this quotation, this question, one could quite easily see a seminal, a final question that fuels all the rest: Why can't we fathom the base-line, first-causal matter of creation? "Why does the universe go to all the bother of existing?"[169]

Such a mind as Hawking's cannot be minimized or taken lightly in any respect. And, it is my personal opinion that such intelligence is not accidental or coincidental. God, in His wisdom, has His own deep reasons for giving mankind such exposure to such deep inquisitiveness and intellect, and we should be able to take it as a springboard into the quest for truth. Indeed, what is it that "breathes fire into the equations and makes a universe for them to describe?"

There are a select few fine scientific minds that follow the path to Jesus Christ as Co-Creator and who understand that He is the Holy Fire that burns there and in all of the creation that so captivates the minds of

168 Copyright @1988, 1996 by Stephen Hawking: Bantam Books, Division of Random House, Inc.
169 Ibid.

247

these scholars. Why the answer comes as such a difficulty to the vast majority of secular-minded science scholars is, to me, a sad thing, because seeing truth in the music of the cosmos is so much more captivating in its reality and color. Did I say seeing "truth in the music"? Yes, and that is precisely what I meant to say: music is heard, but music is also seen. That is the abstract definition of color: the color of music.

There are privileged individuals who have the fascinating gift of synesthesia, which is described as: "a concomitant sensation; especially: a subjective sensation or image of a sense (as of color) other than the one (as of sound) being stimulated."[170] In the 1990's, I was teaching choral music and theory in a community college in Florida. I had a student whose major instrument was the violin. Heather had the gift of synesthesia and spent long times of discussion with me, explaining or describing her ability to "see" music. I believe that she had great fun trying to coax me into being able to do the same thing, but, obviously, it never happened—and still has not been my lot in life. I think I envied her, but I am not sure that she saw her "gift" quite the same way I did.

What it did do for me, however, was to incite my already active curiosity and imagination about the relatedness of all things. If music had "wave lengths," why then, could we not assume with some degree of accuracy that the "wave lengths" of color and light were just as real and relevant? And quantifiable? This was earlier in my process of discovery, and it has been gratifying to me to discover that science has been on the same quest for eons and is now, actually, through the efforts of men like John Stuart Reid[171] and his colleagues, developing the means of analyzing and documenting these phenomena. Here is a quote from one of the "cymascope" websites that may help to clarify the basis of this new science:

170 http://unabridged.merriam-webster.com/unabridged/synesthesia
171 http://www.cymatics.co.uk/author/john/.

"The generic term for the patterns of vibration that occur on the surface of an object when excited by an incident sound is 'modal phenomena,' a field of study that covers everything from vibrations in suspension bridges, to vibrations in body parts of cars, to the effects of sound on the human skeleton and internal organs. In the 1970's this branch of science was named 'cymatics' by Swiss doctor Hans Jenny, a word that derives from the Greek 'kyma,' meaning 'wave' and the inspiration for the name of our CymaScope instrument. The classical view of modal phenomena is that modal patterns form as a consequence of the natural resonant frequencies, or modes, of the object or membrane; current mathematical techniques used to describe this class of phenomena say nothing about the quality of the exciting sound. Musical sounds contain many harmonics so when a circular membrane is excited by a complex musical sound the resulting modal pattern(s) are, naturally, also complex. If we sample a moment from music and analyze it in terms of its fundamental frequency and associated harmonics, and then apply that sample to, say, a circular latex membrane of known elasticity, known diameter and fixed edge, present mathematical techniques cannot predict what pattern will form on the membrane. It appears that no one has attempted to solve this problem, either because no applications for a solution have become evident or because physicists have not seen the importance of mathematically modeling such phenomena. Only the pattern associated with the fundamental frequency can be predicted with any degree of certainty. Thus, for

example, the design of musical instruments remains an art
rather than a science."[172]

The interest in being able to see color when listening to music pre-
dates by far this new emerging science dealing with modal phenomena. I
seldom use Wikipedia as a research tool, but they have done a particularly
good job of making information about this idea accessible:

> "The interest in colored hearing dates back to Greek antiq-
> uity, when philosophers asked if the color (chroia, what
> we now call timbre) of music was a quantifiable quality.
> Isaac Newton proposed that musical tones and color tones
> shared common frequencies, as did Goethe in his book,
> 'Theory of Color.' There is a long history of building
> color organs such as the clavier à lumières on which to
> perform colored music in concert halls.
>
> "The first medical description of 'colored hearing' is in an
> 1812 German thesis by the German physician Sachs. The
> 'father of psychophysics,' Gustav Fechner, reported the
> first empirical survey of colored letter photisms among
> 73 synesthetes in 1876, followed in the 1880s by Francis
> Galton. C.G.Jung refers to 'color hearing' in his Symbols
> of Transformation in 1912. Research into synesthesia pro-
> ceeded briskly in several countries, but due to the diffi-
> culties in measuring subjective experiences and the rise
> of behaviorism, which made the study of any subjective

172 http://cymascope.com/cymascope.html.

experience taboo, synesthesia faded into scientific oblivion between 1930 and 1980."[173]

Perhaps two hundred years ago, the study of the "ethos of music" also fell into scientific disarray and became an archaic study and even fell into complete disuse as a viable intellectual idea, only to be enjoying a resurgence today. It is possible that, while these two fields of knowledge may not be considered "cutting edge" information today, they are beginning to be seen as ancient information which may still be useful in some peripheral way. It has always been a point of great interest to me that the ideas and ruminations of ancient scholars so often enjoy such a renaissance and may even find their way into the mainstream of exciting rediscovery.

These considerations are being explored by way of building a structure of inquiry that will hopefully lead to a deeper consideration about the relatedness of all things and the ways in which all knowledge and all matter can be viewed as emanating from One Center. In the Christo-centric world, John 1:1-3 is the centrality, which names Jesus Christ as the Co-Creator with God the Father and the Holy Spirit, and Colossians 1:17 states unequivocally that all things were created for and by Jesus Christ and in Him all things hold together. In that world of "Christ-centeredness," there is no question Who that One Center is; it is Jesus Christ, Son of God. And if, by use of the sanctified imagination, one can allow oneself to grasp the concept of unity because the design is unified and the Designer is central and holy, governed by the beauty of holiness, which is inherent to the Designer, one could become quite excited by the concept of the sound of the voice of God as being that holy sound, which binds all

173   https://en.wikipedia.org/wiki/Synesthesia.

matter together. And, it is also the template for providing understanding of all sound and the frequencies which identify and codify them. God put all of these mechanical laws into place as part and parcel of all the systems intrinsic to all creation.

UNIVERSE: THE WHOLE COSMIC SYSTEM OF MATTER AND ENERGY OF WHICH EARTH, AND THEREFORE THE HUMAN RACE, IS A PART.[174]

Nothing will challenge and enlighten the mind more than the simple act of reading—**if,** that is, the intent behind the reading is **to learn or to understand.** The simple definition of the word "universe," under the influence of creative or sanctified imagination, can trigger a wealth of knowledge which one may have gathered over a lifetime.

As a not-too-enthusiastic freshman student of general science back in the '40's, I can remember Miss Lauber telling us that energy could neither be created nor destroyed. "How come?" Well, believe it or not, there was no satisfactory answer for a 12-year-old whose favorite question was "Why"?" or "Why not?" And that question remained with me for over half-a-century. It remained until one day when I was pondering the truth of creation and this fairly new idea I had that there was actually "truth in music," **and**—I had to admit to myself that I did not even know yet what I meant by this idea of "truth in music." And besides, what did that have to do with the supposed fact that energy was constant—it could change shape and expression but it **could not be destroyed or created.**

In those 50-odd years of journey, I had learned about string theory. I had learned that Einstein had the firm conviction that there was—somewhere—somehow—a unified theory of everything that would provide the science

---

174   https://www.britannica.com/topic/universe.

of physics with one simple and beautiful set of laws that would explain everything. I had learned that there was a new and emerging science called "cymatics" that had finally reached the place where they, the "cymaticists," were prepared to make the statement that without sound, matter would disintegrate and the universe would be no more. Sound holds everything together. Furthermore, as we have noted earlier, with newly designed methods and equipment, they are beginning to be able to provide empirical evidence to that extent. This could be a revolutionary event. New turns in development always come at the time when most needed.

I had learned to my own personal satisfaction that there absolutely is a thick thread of constancy and relatedness in all things. Not being a scientist, I must admit to being unable to provide the acceptable documentation for this belief, but I can, and am definitely attempting to, provide the impetus for other scholars to become interested, inspired and motivated to follow this train of thought:

> There has been an ongoing notion that in the act of creation, the idea of matter's being created out of "nothing" is recognized in the worlds of theology, science, philosophy, and many other disciplines as well. Gradually, over the ages, a means has finally materialized to give credence to this idea. Cymatics as a new, emerging discipline of knowledge is now willing to name that "nothingness" as sound, vibration and frequency. String theory also supports that idea to the extent that it is no longer just an "idea," but a credible theory.

> In the meantime, the music world has been rapidly moving toward a new understanding of the impact of

music. It has been used in childbirth to relieve pain, music therapy has a growing influence in the world of psychology and psychiatric treatment, cancer patients have benefited immensely from the use of music as a part of their treatment, autism, cerebral palsy—ever so many different conditions have changed the course of development and/or remission by music. I have personally witnessed two victims of Alzheimer's debilitating symptoms finding relief in playing the piano or singing. There is a surfeit of examples that could be given, but here we are—now at a crossroads where we might come to a definitive conclusion in time.

If, indeed, all of creation is a result of vibration, frequency, and sound, why can we not allow our sanctified imaginations to take us to the next level where we might see—or hear—the truth of the matter. What is that vibrating sound, what is the energy that drives it and, beyond that, from where did it emanate?

Remember that Curt Sachs over one hundred years ago, said that "however far back we trace mankind, we fail to find the springing up of music." The Hebrew language contains many hidden thoughts and ideas. Many of the action words in that wonderful living language have the dual meaning of speaking or singing. What if, just what if, in the beginning, as Genesis says, it was the magnificent and eternally powerful energy of the voice of the Singing God that hurled the galaxies and the universes into space and launched just the required amount of singing energy required to create, continue and command everything that is now extant for our study, exploration and joy—and to provide us with a vehicle for praise to that Almighty God?

That would provide, as well, the total explanation for the inability for energy to be created or destroyed. Like God, its Maker, that energy simply is! And it is musical energy! It would also endow us with the perfect landscape for the ultimate truth, the "truth in music," which is God's creative voice singing us into being and continue to hold us and all that we inhabit together simply by the sound of that voice?

The idea in no way detracts from the truth that we have held for eons, that Jesus Christ is the Way, the Truth and the Light, because that truth is expressed by His omniscient musical Being. Truth, like energy, is inviolable. And both belong to God alone.

Take the verities of these last few paragraphs and examine them under the light of knowledge of all things, but especially the knowledge of God and His Word, and see what the result would be. I suspect it would be the illumination of His truth in a slightly new focus—and just maybe that new focus would tie together the loose ends of all other disciplines which have for eons been searching for that one missing thought—a set of simple beautiful laws which would grant us a vision what Einstein intuited was there all the time.

WHEN THE MORNING STARS SANG TOGETHER, AND ALL THE SONS OF GOD SHOUTED FOR JOY. DOES THE COSMOS SING?

God knew something from Day One—and before— that He had great interest in sharing with His created beings—like Job. Just allow your Jana to converse with you about what it would be like to eavesdrop on God and Job or God and Michio Kaku, string theorist extraordinaire, as they discussed the first minutes or hours of the newly created earth. God knew exactly how He had accomplished His great feat of Creativity, but He also knew that mankind could not encompass such a thing in his newly minted mind. As a teacher, early on I learned that one of the finest tools for teaching was to "ask." God is still asking us. The problem is that we are, as finite man, far more interested in proving our own wit and wisdom than we are in actually learning. So why not listen to the stars that God was challenging Job to consider?

NASA is presently engaged in numerous experiments and observations tracking events that are constantly taking place in our galaxy as well as myriads of other galaxies that we can observe. A system of observation called stellar seismology[175] is making huge strides in locating and recording these events and trying to document their various impacts and results. Some of the sounds found and recorded by NASA, are, to say the least, astounding and fascinating, but above and beyond that, the amount of knowledge being accrued will provide us with material for analysis for a long time to come. These radio emissions from galactic sources, especially the sun, are responsible for our earthly auroras and it is undeniable that these happenings are accompanied by incredible sounds which are not unlike what we imagine as "space music." The website science.nasa.

---

175 https://www.youtube.com/watch?v=QYs-gUpmF2E.

gov provides a massive amount of information from which we can learn about this line of research.

Many years before I was ever the least bit aware that I would someday write a book about the miraculous universe in which I live, I was enamored of the Aurora Borealis. I wrote about this in *God's Song*, but I think it is worth repeating here. It was early winter in the Hudson Valley of New York. Walking to choir practice one Thursday night in the nipping cold of late fall, I was suddenly stunned to see a sheer curtain of gossamer green whip across the sky. I had never seen anything like that before, but we had just finished a unit in general science about a phenomenon called the Aurora Borealis. The pictures in the science book were brilliantly colored and the wisping shapes of the colors were fantastic to look at.

This might be the same thing, but it was ever so much more brilliant, and it seemed to be alive. I stopped dead in my tracks. As I stood transfixed and staring, the colors shifted from bright green and turquoise to a brilliant red with deep purple shimmers that came and went. There seemed to be an electric tension in the air, though I would never be able to prove that it was real. Then, just as suddenly, the crimson curtain whipped around again and, like a giant kaleidoscope, it took the crystalline-sharp form of a cathedral dome, with all the points meeting in a sharp focus right over my head. But it still shimmered and whipped in varying hues and tones of the colors it was flaunting at the moment. What was it? I broke into a fast run, headed for the church and safety, for I was afraid that the beauty there in the sky might be a forerunner of my being transported into heaven!

I arrived breathless and frightened at the parsonage door, trying to tell our pastor's wife what I had seen—but it was gone.

At least half a century later, I was web-surfing on my new computer and quite serendipitously stumbled onto an article that said the "Aurora

Borealis sings." Someone in a university far away from where I was had gone to Alaska, filmed an aurora and recorded the electro-magnetic sounds that were assumed to be coming from that display of beauty. Of course, at that time, the entire exploration of the cosmos, space, black holes and all the rest was either very, very new or actually non-existent, and there was pitifully little documented information about what really lay in the hinter-spaces of our universe and galaxies.

Since then, we have been afforded volumes of writings and documentation on these subjects, and the further we go into the study of such things, the more exciting it becomes.

In 2008, Professor Jean-Pierre St. Maurice,[176] Department of Physics, University of Saskatchewan, Canada, became deeply involved in researching the aurora and accompanying disciplines, and he said this:

> "The sounds (of the aurora) are electro-magnetic waves which are heard as noise and emitted by an environment antenna located nearby—When you combine the recent solar activity of the sun and the amazing display of the Aurora Borealis and these strange sounds heard around the world, what is the common factor? Electro-magnetic particles—basically shot from the sun."

Professor St. Maurice and his colleagues have, by their specific research into this phenomenon of electro-magnetic beauty, paved the way for deeper understanding in many other arenas of study of outer space and the strange and unusual events that take place there.

176 https://www.youtube.com/watch?v=8vJI8ijY5p0.

This is profoundly important to me in my search for "truth in music," which has to do with the "truth" of creation, the "truth" in theological reasoning concerning the creation and in making the necessary connections among disciplines. Professor St. Maurice mentions often "connecting the dots" which is an essential, if we are to solve the deep puzzle of the reasonable relatedness in nature, the cosmos, and all things pertinent to life in all of its dizzying multiplicity.

The secret lies in understanding sound as the coalescing, cohesive force, and in understanding that the essence of music is frequency, vibration and pitch/sound—all of which co-exist in every area of our universe.

Where was Job (where were we?) when the foundations of the earth were laid? God asked that question the first time and is still asking that question. I personally believe that science, through the study of space, time and matter, is coming very close to being able to see the answer—or maybe better said—is coming very close to being able to hear the answer and see its holy beauty.

Scene V

Truth in Music from the Beginning to 2016

> Consider this…
>
> *Science now knows that the earth has a mantle of frequency and vibration all around it.*

Music in Man's Mind

"Music is a moral law. It gives soul to the universe, wings to the mind, flight to the imagination, and charm and gaiety to life and to everything." Plato

"Music expresses that which cannot be said and on which it is impossible to be silent." Victor Hugo

"Music is a higher revelation than all wisdom and philosophy." Ludwig van Beethoven

"I see music as fluid architecture." Joni Mitchell

"Music when healthy, is the teacher of perfect order, and when depraved, the teacher of perfect disorder." John Ruskin

"It is cruel, you know, that music should be so beautiful. It has the beauty of loneliness and of pain: of strength and freedom. The beauty of disappointment and never-satisfied love. The cruel beauty of nature and everlasting beauty of monotony." Benjamin Britten

"The most injurious thing that can happen to music is one person's concept." Robert Shaw

261

Music in Gods' Mind

"Break forth into singing, you mountains, O forest, and every tree in it! For the LORD has redeemed Jacob, And glorified Himself in Israel." Isaiah 44:23

"Sing, O heavens! Be joyful, O earth! And break out in singing, O mountains! For the LORD has comforted His people, And will have mercy on His afflicted." Isaiah 49:13

"For you shall go out with joy, And be led out with peace; The mountains and the hills shall break forth into singing before you, And all the trees of the field shall clap their hands." Isaiah 55:12

"The LORD your God in your midst, The Mighty One, will save; He will rejoice over you with gladness, He will quiet you with His love, He will rejoice over you with singing." Zephaniah 3:17

"Let the rivers clap their hands; Let the hills be joyful together before the LORD." Psalm 98:8

It was 1978. It was early summer. It was my first flying lesson. I was 46 and very nervous. Of course, I had prepared myself as well as I knew how by reading the required pages from an FAA manual on basic flight principles and procedures. But even as a total novice, I was acutely aware of one thing: I knew absolutely **nothing** about how this thing worked or what the laws of physics were that would be my death or redemption from here on, and I was really at the mercy of the airplane and those laws that would keep me in the air or not allow me to stay in the air. I assumed that I would soon learn enough about it to be safe, both for myself and for my instructor. But I also knew that I had better do "it" right from the very beginning or I would not last long as a flight student—or a human being for that matter.

Flying is an excellent discipline from which to learn some of the most basic rules of life. There really is no room for error, and the joy that comes from doing it right and getting the required results—continued life and limb—are quite wonderful. It is a shame that all disciplines of life do not have that same acute degree of success or disaster built into them, so that as we go along there is no doubt about the "rightness" or "wrongness" of things. But that caveat would not give us human beings much room or even incentive for personal development and growth. Nevertheless, I am eternally grateful for the peripheral lessons learned from those years of my life as a flight student and, subsequently, as a qualified, certified pilot. Whether I understood it or whether I did not, the laws of flight were set somewhere ages and ages past and all I had to do as an excellent pilot was to abide by them.

It is not so with the problem of music. Music, while it may indeed have incredible peripheral effects on the human psyche, does not present matters of immediate demise or death. However, as Tolkien so powerfully wrote in the *Silmarillion*,[177] the evil god, Melkor, did an incredibly effective job of spoiling the creation of Ea with his evil singing. And, perhaps, this is not a far-fetched notion, and, as we have discussed in *God's Song*,[178] it may be that this very same principle is at work in our own universe and, lacking severe controls on the creation and evolution of music through the ages, our understanding of what it really is and how it was initially intended to be managed as an integral part of the perfect initial intent of creation has been grossly infected by elements of the ungodly. This is an area of conjecture that cannot, on any level, grant satisfactory conclusions, but it does offer a fertile field for some interesting excursions into possibilities—possibilities about the beginnings of what we know as music, and possibilities of the original intention and objective for music in its entirety.

---

177 *The Silmarillion*: J.R.R. Tolkien: Copyright 1999 by Christopher Reuel Tolkien: Random House Publishing Group.

178 *God's Song* 1st edition: Liberty University Press: Lynchburg VA: 2010; *God's Song*, revised 2nd edition, Publishers Solution: Forest, VA: 2017. This book is the "pre-quel" to *God's Opera*.

Before going much further into this discussion, I must repeat a caveat which I have been referring to throughout this quest for "truth in music." The music which will be at the center of our query at this point is not the familiar music of our culture or of the myriads of cultures over the past many hundreds of years. The bedrock of my theory of creation as a musical event, composed and orchestrated by Holy God Himself, is not in any way that I can imagine, an event that can be described by any of the music of man's making. You see, scholars of music for eons of time have been declaring that music is an activity, and, yes, a creation, of man himself, not of God Himself. That completely misses the point, as I see it.

We human beings are bound by our finiteness, whether it is convenient or comfortable for us to admit it or not—we **are** finite and that limits to a very large degree our understanding of anything other than the world that we can see, touch and manipulate. Therefore, our prime mode of understanding is also finite, and we cannot grasp the eternal or the spiritual through any other filter but our finiteness. That makes us immensely vulnerable to the belief that man, after all, is the center, the creator and the builder, and beyond that, to where we think the idea of God as omniscient, omnipresent and beyond all understanding is all but impossible, as well as humbling. This leads to a pervasive need to see man as the center of all knowledge and accomplishment, and it leads to the necessity for empirical proofs for everything we see and do.

The antithesis of that mode of acceptance is faith. That is precisely why, in the chapters on faith, I tried to make the point that faith is a quality found in every facet of life and even the profound thinking of the scientist must be undergirded by faith in **something**.

What is it, then, that we should be learning from contrasting what man thinks about music as opposed to what God **may** think about music and what His original intent was **for** music. You see, without faith, it is impossible to know, to see, or to please this God Who made all from nothing but the sound (the vibration) of His voice.

MUSIC IN MAN'S MIND

Music **is** a moral law, I think. I would like to know what Plato's reason was for making such a statement. My reason(s) for agreeing with him are probably quite different from what he was thinking, but it would be interesting to find out. After more than forty years of journeying to find that elusive eminence which some of us call "truth in music," I have finally begun to see spiritual verities which I could not see before, and more and more as I travel, those verities point to the reality of music as intrinsic to the creation saga.

Victor Hugo, nearly two hundred years ago, understood the magnanimity and overarching power of music and was awed by it. He did not understand it, however, I am sure.

Certainly Beethoven struggled his entire life with the verities of musical truth—and spiritual truth as well, I think. I believe some part of his soul, his entire being understood that there was a perfection to be sought after and that music had **something** to do with that, but he never ascended to that height where he could understand God as intrinsic to that truth. Or maybe he did, and just maybe that is why he suffered so much in just living. He certainly was able to express truth in works such as the *Missa solemnis*!

We could go on and on pondering the thoughts of Joni Mitchell, John Ruskin, Benjamin Britten and many more, but I believe that Robert Shaw teetered on the brink of final comprehension when he said, "The most injurious thing that can happen to music is one person's concept."

How true! And precisely the reasoning behind my beginning the original quest for "truth in music" nearly half a century ago. Musicians are notorious for their individual preferences and are not shy about stating them either. And, so, thousands of years ago, we—that is, we musicians—launched on a millennial search for the "right kind of music," the "real ethos of music," the answers to a thousand questions about the need for, uses of, justification of and explanation of that music which any one of us deemed most imperative. It has been an unsolved quest, and as the plethora of music of all genre burgeoned, the unanswered questions have grown exponentially as well.

The hazy thought that haunted me for so many years was simply that there had to be some gravely important raison d'etre` for the existence of music in the first place and some evasive starting point that would might even coincide with Einstein's "one set of beautiful laws" which would provide such insights and answers.

If sound is the cohesive factor that science is now laying claim to in all creation, music as the raw material of the spoken or sung creative act could become more clearly understood.

MUSIC IN GOD'S MIND

The Scriptures, especially the Old Testament, are filled from Genesis to Malachi with so many references to the hills, trees, mountains, springs, forests, rivers and rocks, and even the heavens, breaking forth into singing, clapping their hands for joy and generally making sounds and singing songs in praise and joy. In Isaiah 49:13, there is a particularly interesting command made. Take special note that the admonition is not made in the form of a suggestion or request. It is a statement of demand requiring **action:**

Act 4 — Scene V

"Sing, O heavens! Be joyful, O earth! And break out in singing!"

Almost without exception these statements are followed by some clear expression of praise to Almighty God for some wonderful benefit He has provided. There seems to be a greater acceptance of the idea that, indeed, the universe and all that is in it could have been created out of nothing, just as the Scripture has said for time immemorial. *"And God said, ...and it was so."* Only the energy of the vibration of God's voice: *"and it was so."*

The most captivating suggestion is in the Book of Zephaniah where it is clearly stated, *"The Lord God in your midst ... will rejoice over you with singing."* When I first read in the Old Testament that God actually does sing, the rest of the puzzle of creation and its mechanics began to fall into perfect place. Of course, this entire theory is based on the premise that Scripture is not fantasy—but fact.

But before one condemns the possibility, one should spend time examining the idea under the microscope of reason, thoughtfulness and the scientific method. It is becoming a more and more sound conviction that if we read and examine Scripture with an intellectual and scientific inclination, most of the mysteries that plague understanding would vanish.

Earlier in this book, we spent time with the ancient Hebrew language. One of the intriguing things about this language is its living quality. I remember my professor telling us that Hebrew was so difficult because it depended so deeply upon context for clarity—and how can context provide clarity when context changes almost concomitant with the number of venues, issues or people being considered. Another puzzling thing about the language is the fact that there are no superlative descriptive words.

Superlative ideas are understood by how many times the idea is repeated. With this in mind, consider the repetitive references to singing, playing on the harp or flute, praising God, and making a joyful noise. Also, consider the intriguing and over-abundant reference to the things of nature—trees, mountains, stones, heavens and seas—clapping their hands or singing praises. Do the things of nature actually do that?

It would seem that now, science is beginning to know that this is the case—the planets sing, the earth itself has a mantle of frequency and vibration all around it called the Schuman Cavity or Schuman Resonance. And we are being told that we have an undeniable and irrevocable connection to the planet on which we live called "a frequency."[179] James Russell has produced an incredible film noted below which I hope every reader will avail himself of.

Armed with this fairly new knowledge, it is becoming far more reasonable to read the predictions of the Old Testament in the form of God's Words about music (frequency, vibration and sound—musical sound) found in the veiled and beguiling verses about music and beauty and finally be able to, with your sanctified imagination, envision the truth that all creation vibrates with song in praise to Almighty God.

To go one further step and begin to understand what this "music of God" is, becomes a highly challenging process. Furthermore, out of this growing understanding, comes a deeper understanding of what the music that we human beings can enjoy and perform actually is and from where it emanates.

179 http://www.earthbreathing.co.uk/sr.htm.

$\mathscr{S}cene$ *VI*

TRUTH IN MUSIC REALIZED? DA CAPO

> *Consider this…*
>
> ## Musical vibrations, pitches and sound are evident in all nature and in the cosmos at large.

Throughout the writing of this book, I have attempted to open each chapter with nuggets of thought, mine or otherwise, in order to inspire creative thoughts in the reader's mind—

Typical of the Baroque era in keyboard music was the "figured bass:" "a bass line with the intended harmonies indicated by figures (intervals designated by numbers) rather than written out as chords, typical of continuo parts in baroque music."[180]

It then became incumbent upon the harpsichordist or organist to "realize" (to fill out in complete harmonies, complete with passing tones and other harmonic technicalities) the bass part, which in turn, supported the entire composition and gave depth, rhythm and a solid harmony to the whole. If the "figured bass" were not properly "realized," the entire performance would be inferior and flawed.

I believe that creation, in the beginning, was, in fact, a musical event, which has left all of the universe and everything in it, fully crafted from

---

180   https://www.google.com/#q=figured+bass.

and supported by musical elements—perhaps not as we know musical elements—but the foundational elements of music are basically inviolable. And now that it is time to take the notions and notations of multiplied years and put them into some semblance of a " figured bass," to be realized, (if, indeed, that metaphor is sound enough to use) I, as the presenter of this theory, "realize" that this writing will never be known as a completed work. I can begin to understand the mind of Bach, who undoubtedly could not suspend his work until it ended naturally at his death. And maybe there is an intrinsic idea here, that all things of eternal nature, including beauty and the beauty of holiness, really have no end. They go on until the time at which the Creator God sees fit to place the period at the end of this age—and then—

Why is it so important to justify the idea of a musical creation saga? Is there an underlying theme that has the potential of providing clarification and/or answers to so many other enigmatic theories and postulates that remain fixed or unanswered in our system of knowledge—knowledge of all things? I believe that there is—and after so many years of searching, I am increasingly convinced that music in some form, **that we may or may not know**, is, indeed, that theme.

This idea is all but impossible to frame in words. It is true that over the ages, scholars and students alike have pondered over the arts—what they are, why they are and what they actually accomplish for mankind. I have always been offended to some degree or other by the notion that music, for instance, along with the other arts, I suppose, enhances the quality of life and is useful for the purpose of providing a more worthy use of leisure time as one listens, performs or ponders over it. There is much more to the existence of music than that; I believe that musical understanding holds the key to all knowledge. Perhaps that is why the surface or superficial approach to music leaves so much to be desired.

Act 4 — Scene VI

Try teaching the average music appreciation class with the average approach limited to such thinking and then watch the faces of your students as they struggle to hide their boredom. On the other hand, try speaking to the average music appreciation class about the frequencies of electromagnetic waves in the flashing, shimmering waves of light in the Aurora Borealis and how that translates to the living, singing miracle of music in the cosmos, and then, suggest that Gustav Holst may have been trying to capture the miracle of that very phenomenon of cosmic music when he wrote *The Planets*. It suddenly gives life to the whole idea of "music appreciation."

You see, it has occurred to me that we approach this whole idea of what music is from the wrong end of the spectrum—it is not, by any means, simply a man-made creative activity for the sole purpose of human elevation and enhancement. And, it also is much more than simply one of the marvelous gifts of God given to His creatures for their good and enjoyment. Rather, I believe that its genesis, the origin of music and the other arts, comes from God as a dim, smoky-mirrored replica of His unattainable and ineffable creative power. It is an intrinsic part of the concept that we are made in the image of God. Perhaps, the concept of **musicality** as one of the attributes of God is really not too foreign to consider.

The inherent problem is that, whether or not we are willing to admit it, our human concepts are bounded by the very nature of our humanness and we are innately limited to the knowledge of things which are within our scope of knowing and our span of awareness. Therefore, when we talk about music, the first and maybe, the only, thing that we can intelligently discuss, is music as we know it to be, and that discussion is extremely narrow and depends almost totally on the preferences and level of knowledge held by the people involved in that discussion. We have no frame of reference for what God's music may be like nor what His musical intents

271

may be or might have been. Most assuredly, music from God's vantage point would be constrained to be far more deep and complex and essential than anything we may know.

The idea of "truth in music" is not new. But the differences in such a concept from the point of view of mankind would be very far separated from the concept of "truth in music" from the eternal point of view. Perhaps at this point I should make another caveat. When I speak of "truth in music," I am referring to the truth of creation, not the concept of Scriptural truth that refers implicitly to the atonement and other foundational doctrines of the church. The idea of "truth in music" in no way contradicts any of those doctrines. It may even expand upon them or enhance the expression of them, but this entire journey of discovery has to do, basically, with creation and the genesis of the universe and all that is in it.

I was talking with a friend one day after church, and we were discussing the phenomenon of the Schumann Cavity, that space between the earth's outer crust and the outer reaches of the ionosphere and the electromagnetic tension between them. The earth's pulse-rate, or the vibrational exchange rate, according to scientific research is 7.83 Hz, which equates to 7.83 vibrations per second. The pulse rate of the ionosphere is also 7.83 Hz. The electromagnetic wave-length of the human brain in its alpha—working, awake state—is 7.83 Hz. Coincidence? Serendipity? I think not. Divine Design? Of course! Even so, what does that have to do with "truth in music"? Or with our concept of creation as a whole?

FOUNDATIONAL MUSICAL ELEMENTS

Within the pages of this book, we have perused four supposedly unrelated and diverse disciplines—or areas of intellectual consideration—which are language, science, theology and music. On the face of it, these are disciplines that are not directly or closely related. But hovering above all of the research and ideas explored over the last forty years is a common thread—there is a definite relatedness in all of creation and in all of the bodies of knowledge as we know them.

The idea of relatedness and design has energized scholars from all walks over hundreds of years: Pythagoras, Archimedes, Aristotle, Copernicus, Galileo, Kepler, Newton, Fibonacci, Roentgen, Einstein, Faraday, Kaku, Greene, actually too many to count. Recently, even new disciplines have been emerging which will hopefully give new energy to the search for truth in such a way that the relatedness can be proven.

As noted previously, this new discipline is called cymatics and has emerged basically from the UK. This group of scholars has successfully designed and built an instrument called a cymascope which successfully records images of vibrations created by frequencies fed into it. These images can be photographed and analyzed. From the cymatics website:

> "Vibration underpins all matter in the universe. No matter can exist without vibration.

> "The CymaScope is the first scientific instrument that can give a visual image of sound and vibration in ways previously hidden from view. When the microscope and telescope were invented they opened vistas on realms that were not even suspected to exist. The CymaScope holds the same potential as the microscope and telescope and its applications are beginning to touch a broad range of human endeavor. This website is constantly updated so please visit us every few weeks for the very latest in

cymatics research. Our ongoing mission is to support and advance the emergent science of cymatics."[181]

This is, in my estimation, one of the most exciting findings of the 21st century, and it could easily be the discovery that completely reverses the direction of research on the genesis of our worlds.

The essence of music is vibration, frequency and pitch. Science is beginning to tell us that the essence of all matter is vibration, frequency and pitch. I have not, up to this point, seen any printed material that is outspokenly making the claim that these two ideas are intertwined or related, but with just a slight margin of allowance for sanctified, imaginative thinking, it becomes easier and easier to connect those dots. When that connection is finally made, it will be a small step to the realization of this whole matter—that creation happened as the result of sound—what sound? That creation is held together by sound. What sound? That there is a direct relationship between sound and matter.

And this is the precise point at which the divine realization dawns. *"And God said ...and it was so."* The sound of the voice of God in the Person of Jesus Christ, the Co-Creator, and it is my conclusion, drawn after so many years of searching and researching, that that specific sound was musical in form and format—as the Hebrew language draws into our attention span, the word *dabar*—"to, speak, to sing, to create, to command."

What other explanation is there for the fact that musical frequencies, musical vibrations, musical pitches and sounds of all genre are so evident in all of nature and the cosmos at large? And why are they so orderly so obviously consistent in so many differing venues?

181　http://www.cymascope.com/.

Act 4 — Scene VI

REPRISE

In Conclusion—*Return To The Road Less Traveled By*
"I took the road less traveled by, And that has made all the difference."

- Genesis 1:1: *"In the beginning God created the heavens and the earth."*

- Genesis 1:3: *"And God said, and there was...."*

- Psalm 33:6: *"By the word of the Lord the heavens were made, and all their host by the breath of His mouth."*

- Hebrews 11:3: *"By faith we understand that the universe was framed by the Word of God, so that things that are seen were not made out of things which are visible."*

- John 1:1-3: *"In the beginning was the Word, and the Word was with God and the Word was God. He was in the beginning with God. All things were created through Him and without Him nothing was created that was created."*

- Colossians 1:16: *"For by Him all things were created that are in the heavens and the earth, visible and invisible ...All things were created by Him and for Him. And in Him, **all things hold together**."*

How incredibly overwhelming to learn that science now, in the voice of the newly-emerging discipline of cymatics, is making the unequivocal claim that without sound, all matter would cease to exist. And what could that sound possibly be? The voice of God...

275

About the Author

ettie Harris is retired from a long and colorful career as a teacher both of music and aviation. She received her early musical education from private teachers at the military academy at West Point, New York, but it was later in life when, at age forty, she began her formal training at Westminster Choir College and Montclair State University in New Jersey to become a degreed teacher. During her years at Westminster, she sang in the renowned Westminster Choir, singing in many well-known concert halls and serving as a member of the choir in residence at the Spoleto, Italy, Festival of Two Worlds. She became a licensed FAA instructor pilot in 1981.

Throughout her myriad experiences as teacher and performer, Harris maintained an active pursuit of what she called "truth in music," which embraces not only her fervent faith in God, but also the raison d'etre of music as an integral part of our entire human experience.

Harris now lives with her husband in Forest, Virginia, where she is actively pursuing her research and writing.

CPSIA information can be obtained
at www.ICGtesting.com
Printed in the USA
FSOW03n0008181117
41254FS